PARALLEL COMPUTATION SYSTEMS FOR ROBOTICS
Algorithms and Architectures

WORLD SCIENTIFIC SERIES IN ROBOTICS AND AUTOMATED SYSTEMS

Editor-in-Charge: Prof T M Husband
(*Vice Chancellor, University of Salford*)

Vol. 1: Genetic Algorithms and Robotics — A Heuristic Strategy for Optimization
(*Y Davidor*)

Vol. 2: Parallel Computation Systems for Robotics: Algorithms and Architectures
(*Eds. A Bejczy and A Fijany*)

Vol. 4: Computer Vision, Models and Inspection (*A D Marshall and R R Martin*)

Forthcoming volumes:

Vol. 3: Intelligent Robotic Planning Systems (*P C-Y Sheu*)

Vol. 5: Advanced Tactile Sensing for Robotics (*Ed. H R Nicholls*)

Vol. 6: Intelligent Control: Aspects of Fuzzy Logic and Neural Nets (*C J Harris*)

Vol. 7: Visual Servoing (*Ed. K Hashimoto*)

Vol. 8: Modelling and Simulation of Robot Manipulators: A Parallel Processing
Approach (*A Y Zomaya*)

World Scientific Series in Robotics and Automated Systems – Vol. 2

PARALLEL COMPUTATION SYSTEMS FOR ROBOTICS
Algorithms and Architectures

edited by

A. Fijany and A. Bejczy

Jet Propulsion Laboratory
California Institute of Technology
Pasadena, CA 91109, USA

World Scientific
Singapore • New Jersey • London • Hong Kong

Published by

World Scientific Publishing Co. Pte. Ltd.

P O Box 128, Farrer Road, Singapore 9128

USA office: Suite 1B, 1060 Main Street, River Edge, NJ 07661

UK office: 73 Lynton Mead, Totteridge, London N20 8DH

Library of Congress Cataloging-in-Publication Data

Parallel computation systems for robotics : algorithms and architectures / [edited by] A. Fijany and A. Bejczy.
p. cm. -- (World scientific series in robotics and automated systems : vol. 2)
Includes bibliographical references.
ISBN 9810206631
1. Robots -- Control systems. 2. Robots -- Programming. 3. Parallel processing (Electronic computers) I. Fijany, A. II. Bejczy, A. III. Series.
TJ211.35.P37 1992
629.8'92--dc20 92-19670
CIP

First published 1992
First reprint 1994

Printed in Singapore by JBW Printers & Binders Pte. Ltd.

INTRODUCTION

Modern robotics profoundly relies on computational capabilities: robot action planning, trajectory generation, sensory data and information processing, control, and simulation all involve computation-intensive algorithms which must be processed in real-time or near-real-time. Advancements in application of robotics call for practical computing architectures with substantial computing power.

The conventional serial processors, despite significant improvements in their processing speed, are far from providing sufficient computing capability for advanced robotic systems. Furthermore, like in many other scientific fields, the computational requirement in robotics applications grows much faster than the processing speed of serial processors, leading to an ever increasing gap between the required and the available computing power. Fortunately, the advent of VLSI technology offers a unique opportunity to fill this gap through parallel computation. Parallel processing algorithms and architectures can be designed and implemented to provide significant speedup in various robotic computations at a reasonably low cost.

In this book we concentrate on computational problems arising in real-time control and simulation of robot manipulators. The implementation of advanced control schemes and dynamic simulation of robot arm motion require the solution of a class of kinematics and dynamics problems. Some of the other computational problems, e.g., task planning and sensory data and information processing, are also generic and important in robotics. But the class of real-time control and simulation problems is more specific to robotic applications since they reflect the kinematic and dynamic behavior of the specific mechanical structure of a robot manipulator. A few words about the computational aspects of this class of problems.

During the past fifteen years or so considerable effort was devoted to the development of efficient serial algorithms for this class of problems. As a result of this effort, we have today asymptotically optimal serial algorithms for these problems. In fact, it seems unlikely that any but minor improvement in the efficiency of the serial algorithms can be achieved. An illustrative example is the computation of manipulator inverse and forward dynamics. The first algorithm which synthesized the solution of both inverse and forward dynamics was of $O(n^4)$ complexity with n being the problem size, i.e., the number of Degree-Of-Freedom (DOF) of a robot arm. The prohibitive computation cost of this algorithm made its real-time implementation, even for a robot arm with a small number of DOF impossible. Today, there exist recursive $O(n)$ algorithms for serial computation of both problems.

A robot arm, however, has a fast dynamics which implies that the computation for real-time control and simulation should be performed more frequently and in a shorter time interval than even the best serial computation permits. Motivated by this need, researchers in the past few years developed efficient parallel algorithms and architectures to achieve the required computing power. The parallel algorithms for each specific problem together with the associated architectural requirements are well understood today. We now have parallel algorithms which achieve the time lower bound of $O(\log n)+O(1)$ in solving robotic computational problems with asymptotically minimum number of processors.

A specifically illuminating case is the forward dynamics problem. Until recently it was believed that the time and processors bound in its computation are related to those in linear system solution, that is, it can be computed in a time of $O(\log^2 n)$ with $O(n^3)$ processors. But it has been shown recently that a new $O(n)$ algorithm, though less efficient than the previous $O(n)$ algorithms for serial computation, can be parallelized to solve the problem in $O(\log n)$ steps with $O(n)$ processors which indicates optimal bounds on both time and processors. The fact that there exist $O(\log n)$ parallel algorithms for computation of this class of kinematic and dynamic problems reflects the recursive mechanical structure of the robot arm. That mechanical structure allows the solution of these problems to be cast as the computation of a set of first-order and second-order linear recurrences which can be solved in $O(\log n)$ steps with $O(n)$ processors.

The first four chapters of this book are devoted to parallel computational algorithms for real-time control and simulation of robot arms. In particular, the first chapter handles parallel algorithms and architectures for computing the inverse dynamics of an n-link robot arm. The developed algorithm is based on the Newton-Euler equations and can be implemented on a SIMD computer with "P" processors to achieve the time lower bound in the computation. The second chapter revisits the computation of the robot arm inertia matrix and proposes a new computational model in the form of a "non-strict computation" of the inertia matrix. This model leads to a highly parallel algorithm suitable for implementation on a multiprocessor architecture. The third chapter introduces a parallel processing scheme using the static multiprocessor scheduling algorithms for robot arm dynamic control computation based on the Newton-Euler equations. It has actually been implemented and evaluated on a seven processor elements architecture. The fourth chapter takes a multilevel approach to parallel computation of the robot arm dynamic control and simulation problems. It exploits parallelism at several computational levels by the use of a specifically designed, interconnected and synchronized "Parallel Computing Engine" (PCE) where each processor of the PCE is a SIMD parallel architecture. Several implementation examples have been worked out to show the power of this multilevel parallel architecture for robotic computations.

The fifth chapter of this book considers trajectory computational schemes for single and multiple robot arm systems. It proposes the use of a multiprocessor robot controller for motion coordination and communication with multiple robotic devices and presents a new approach to accomplish real-time task scheduling. Implementation examples are worked out for the proposed multiprocessor controller to illustrate its efficiency in real-time control.

The sixth chapter of this book takes an insider's look at the biologically inspired, computational and information processing paradigm: neural networks. It presents a repertoire of adaptive neural learning algorithms which were exploited to provide an enabling core for addressing a fundamental robotics problem: design of autonomous robots for operating in unstructured and unpredictable environments.

Each chapter of this book has been prepared by different authors based on their work and expertise in this field. The editors gratefully acknowledge their cooperation in preparing the final manuscript of this book.

Antal K. Bejczy and Amir Fijany
Editors

CONTENTS

CHAPTER 1

PARALLEL ALGORITHMS AND ARCHITECTURES FOR INVERSE DYNAMICS COMPUTATION

C. S. George Lee

School of Electrical Engineering
Purdue University
West Lafayette, Indiana 47907

ABSTRACT

This chapter describes efficient parallel algorithms and architectures for computing the inverse dynamics of an n-link robot manipulator. First, based on a functional decomposition of the equations of motion of a manipulator, computational flows of the Lagrange-Euler and the Newton-Euler equations of motion are examined, and architectures tuned to the functional flow of the two dynamic models are also examined. An efficient architecture which tunes to the functional flow of the Newton-Euler equations of motion is proposed. This controller architecture can best be described as a macro level pipeline, with parallelism within elements of the pipeline. The pipeline is designed to take maximum benefit of the serial nature of the Newton-Euler equations of motion. Next, the time lower bound of computing the inverse dynamics of an n-link robot manipulator parallelly using p processors is shown to be $O(k_1 \lceil n/p \rceil + k_2 \lceil \log_2 p \rceil)$, where k_1 and k_2 are constants. An efficient parallel algorithm for computing the inverse dynamics using the Newton-Euler equations of motion was developed to be implemented on an SIMD computer with p processors to achieve the time lower bound. When $p = n$, the proposed parallel algorithm achieves the time lower bound of $O(\lceil \log_2 n \rceil)$. The proposed p-fold parallel algorithm can be best described as consisting of p-parallel blocks with pipelined elements within each parallel block. A modified inverse perfect shuffle interconnection scheme was suggested to interconnect the p processors. Furthermore, the proposed parallel algorithm is susceptible to a systolic pipelined architecture, requiring three floating-point operations per complete set of joint torques.

This work was supported in part by the National Science Foundation under Grant CDR 8803017 to the Engineering Research Center for Intelligent Manufacturing Systems.

Any opinions, findings, and conclusions or recommendations expressed in this article are those of the author and do not necessarily reflect the view of the funding agency.

1. INTRODUCTION

Robot manipulators are highly nonlinear systems and the design and analysis of their motion control require the development of efficient dynamic equations. Two competing approaches are developed for handling the mathematical complexities involved in the dynamic model of robot manipulators. In the first approach, the emphasis is on the Lagrangian formulation of explicit state equations for manipulator dynamics [1,2], expressing the relationship between the generalized forces/torques and the generalized coordinates with the system parameters explicit in the state equations. This is motivated by the growing interest in applying advanced control theory to robot manipulators. Unfortunately, the computation of joint torques for servoing the robot manipulator from these state equations is a computational intensive task with a computational order of $O(n^4)$, where n is the number of degrees-of-freedom of the manipulator. Furthermore, these state equations exhibit too many insignificant details of dynamic characteristics of the manipulator. In the second approach, the emphasis focuses on the Newton-Euler formulation [3,4] of the dynamic model in an efficient recursive inverse dynamics form for computing the required generalized forces/torques for a given set of generalized coordinates, their time derivatives and physical and geometric parameters of the robot arm. To compute the joint torques, the Newton-Euler equations of motion have a computational complexity order of $O(n)$ relationship to the number of degrees-of-freedom of the manipulator, when they are computed on a uniprocessor computer. Solving for all the joint torques/forces using the recursive Newton-Euler equations of motion represents the most efficient method at the time of writing this chapter.

The dynamic model of a robot manipulator is useful for 1) designing and deriving an appropriate controller for controlling the robot arm motion to achieve a desired performance, and 2) real-time simulation of robot arm motion. In this chapter, we shall concentrate on the computational problem involved in the robot arm control, in particular the computation of the inverse dynamics problem.

Manipulator motion control is usually specified in terms of the path traveled by the manipulator hand in Cartesian coordinates or in joint-variable coordinates of the manipulator [2]. For industrial robots, the joint motors of the manipulator must furnish sufficient torques so that the manipulator hand will travel along the planned path. Otherwise, the motion of each joint may speed up or slow down, resulting in undesirable deviations of the manipulator hand from the desired path. To overcome this problem, the manipulator control problem requires the efficient computation of the generalized forces/torques or the joint torques†, from an appropriate manipulator dynamics model, to drive all the joint motors in real time for each set point on a planned trajectory.

† By joint torques we mean joint forces for prismatic joints and joint torques for revolute joints.

At present, various control schemes have been discussed in the literature, which include, for example, resolved motion rate control [5], "computed-torque" technique [6], resolved acceleration control [7], nonlinear decoupled control [8], model-based control [9], and adaptive controls [10-14], etc. In most cases, the control schemes involve the computation of the required generalized forces/torques, from an appropriate manipulator dynamics model, using the measured data of displacements and velocities of all the joints, and the accelerations computed from some justifiable formulae or approximations. Obviously, the execution time for computing the generalized forces partially determines the feasibility of implementing the control scheme in real time. Furthermore, in order to achieve fast convergence of the control algorithm, a sampling rate of no less than 60 *Hz* is preferable because the mechanical resonant frequency of most industrial manipulators is around 5–10 *Hz*. For direct-drive robots, a higher sampling rate is required.

One of the basic control schemes is the computed-torque technique based on the Euler-Lagrange (EL) [6,15,16] or the Newton-Euler (NE) equations of motion [17-19]. Basically, the computed-torque technique is a feedforward control with feedforward and feedback components. The feedforward component computes the nominal joint torques along the desired nominal trajectory to compensate for the interaction forces among all the various joints; the feedback component computes the necessary correction torques to compensate any deviations from the desired nominal trajectory. It assumes that one can accurately compute the counterparts of the dynamic coefficients, $\mathbf{D}(\mathbf{q})$, $\mathbf{h}(\mathbf{q}, \dot{\mathbf{q}})$, and $\mathbf{c}(\mathbf{q})$, in the EL equations of motion (see Appendix A, Eqs. (A-3)-(A-5)) in order to minimize the nonlinear coupling effects among the joints, and use a proportional-plus-derivative control to servo the joint motors. Thus the computed-torque control law has the following structure:

$$\boldsymbol{\tau}(t) = \mathbf{D}(\mathbf{q})[\,\ddot{\mathbf{q}}^d(t) + \mathbf{K}_v(\dot{\mathbf{q}}^d(t) - \dot{\mathbf{q}}(t)) + \mathbf{K}_p(\mathbf{q}^d(t) - \mathbf{q}(t))\,] + \mathbf{h}(\mathbf{q}, \dot{\mathbf{q}}) + \mathbf{c}(\mathbf{q}) \tag{1}$$

$$= \mathbf{D}(\mathbf{q})[\,\ddot{\mathbf{q}}^d(t) + \mathbf{K}_v\dot{\mathbf{e}}(t) + \mathbf{K}_p\mathbf{e}(t))\,] + \mathbf{h}(\mathbf{q}, \dot{\mathbf{q}}) + \mathbf{c}(\mathbf{q})$$

where $\mathbf{K}_v$ and $\mathbf{K}_p$ are $n \times n$ derivative and position feedback gain matrices respectively, and $\mathbf{q}^d(t)$, $\dot{\mathbf{q}}^d(t)$, and $\ddot{\mathbf{q}}^d(t)$ are, respectively, the desired joint position, velocity, and acceleration at time t, and $\mathbf{e}(t)$ and $\dot{\mathbf{e}}(t)$ are, respectively, the joint position and joint velocity errors of the manipulator at time t. Controlling the manipulator to follow a desired trajectory reduces to computing the control equation (Eq. (1)) in every sampling period of time. If one can compute Eq. (1) as fast as possible, this will greatly reduce the degradation of the controller and make the manipulator control system more robust to parameter variation. If we substitute the following joint acceleration (i.e., the terms inside the bracket in Eq. (1))

$$\ddot{\mathbf{q}}(t) = \ddot{\mathbf{q}}^d(t) + \mathbf{K}_v\dot{\mathbf{e}}(t) + \mathbf{K}_p\mathbf{e}(t)) \tag{2}$$

into the complete EL equations of motion in Eq. (A-1) (see Appendix A), the resultant computed torques from the EL equations of motion are equivalent to the joint torques computed from Eq. (1). Thus, computation of the control law in Eq. (1) reduces to how fast one can compute the EL equations of motion with the joint acceleration set as in Eq. (2). Unfortunately, the computation of the control law in Eq. (1) based on the complete EL equations of motion is very inefficient with an order $O(n^4)$ relationship to the number of degrees of freedom of a manipulator. As a result, Paul [20] concluded that real-time, closed-loop control is very difficult. For this reason, simplified and approximate dynamic models were used for controlling the manipulator. Unfortunately, these simplified dynamic models confine the manipulators to moving at reduced speeds where the effects of the coupling dynamic terms are negligible.

Another approach is to compute the inverse plant dynamics using the NE equations of motion. The control law in Eq. (1) can be computed using the recursive NE equations of motion. Similar to Eq. (2), the recursive computed-torque control law can be obtained by substituting the following joint acceleration for joint i, $i = 1, 2, \cdots, n$,

$$\ddot{q}_i(t) = \ddot{q}_i^d(t) + \sum_{j=1}^{n} K_v^{ij}\,(\dot{q}_j^d(t) - \dot{q}_j(t)) + \sum_{j=1}^{n} K_p^{ij}\,(q_j^d(t) - q_j(t)) \tag{3}$$

$$= \ddot{q}_i^d(t) + \sum_{j=1}^{n} K_v^{ij}\,\dot{e}_j(t) + \sum_{j=1}^{n} K_p^{ij}\,e_j(t)$$

into the NE equations of motion (see Appendix B) to obtain the necessary joint torque for each actuator. K_v^{ij} and K_p^{ij} are, respectively, the derivative and position feedback gains for joint i, and $\dot{e}_i(t)$ and $e_j(t)$ are, respectively, the joint position and joint velocity errors of joint i. The above recursive control law is a proportional-plus-derivative control and has the effect of compensating the inertial loading, coupling effects, and the gravity loading of the links. In order to achieve a "critically-damped" system for each joint subsystem, the feedback gains K_v^{ij} and K_p^{ij} can be chosen as suggested in Paul [20] and Luh [21]. One of the main drawbacks of this control technique is that the convergence of the position error vector depends on the dynamic coefficients, $\mathbf{D}(\mathbf{q})$, $\mathbf{h}(\mathbf{q}, \dot{\mathbf{q}})$, and $\mathbf{c}(\mathbf{q})$, and the position and velocity feedback gain matrices $\mathbf{K}_p$ and $\mathbf{K}_v$ [20,21].

The computation of joint torques from the NE equations of motion is the most efficient and has been shown to possess the time lower bound of $O(n)$, when the computation is performed in a uniprocessor computer [22]. It is unlikely that further substantial improvements in computational efficiency can be achieved, since the recursive NE equations are efficiently computing the minimum information needed to compute the generalized forces/torques: angular velocity, angular and linear accelerations, and joint forces and torques. Nevertheless, some improvements could be achieved by taking advantage of particular computation structures [23], customized algorithms/architectures for specific manipulators [24-26], parallel computations [22,27,28], and scheduling and mapping algorithms for multiprocessor systems [29-34].

With the advent of advanced microprocessors and VLSI technology, the rapid decrease in computational costs, the reduced power consumption and physical size, and the increase in computational power suggest that further improvement in the computational efficiency of the NE equations of motion for real-time control of robot manipulators can be achieved by:

(1) Designing a multiprocessor-based controller which is based on an architecture, tuned to the functional flow of the NE equations of motion, to meet the required computation rate and cost for real-time control of robot manipulators. This cost-effective architecture, using off-the-shelf microprocessors, can best be described as a macro-level pipeline to take maximum benefit of the serial nature of the NE equations of motion, with parallelism within elements of the pipeline and can yield a consecutive set of joint torques in about 3 milliseconds (using Motorola MC68020 microprocessors). This is discussed in Section 2.

(2) Designing efficient parallel algorithms for computing the NE equations of motion to be run on single-instruction-stream multiple-data-stream (SIMD) computers to achieve the time lower bound of computing the joint torques. This is discussed in Section 3.

(3) Designing efficient scheduling or mapping algorithms for transforming the NE equations of motion into a highly parallel form which in turn can be scheduled or mapped on a multiprocessor system. This is beyond the scope of this chapter. Details can be found in [33-34].

In this chapter, we shall first explore a cost-effective, multiprocessor-based controller, using off-the-shelf microprocessors, for the computation of the NE equations of motion for real-time control of robot manipulators. Next, we shall prove that the time lower bound of using p processors, $1 \le p \le n$, to compute the inverse dynamics of a manipulator, based on the NE equations of motion, is of order $O(k_1 \lceil n/p \rceil + k_2 \lceil \log_2 p \rceil)$, where k_1 and k_2 are specified constants. If $p = n$, then the shortest parallel computing time is not lower than $O(\lceil \log_2 n \rceil)$. We shall then explore efficient parallel algorithms for computing the NE equations of motion in SIMD machines to achieve the time lower bound of computing the joint torques.

2. MULTIPROCESSOR-BASED CONTROLLER

Recent research efforts are being directed toward designing multiprocessor-based controllers based on a dynamic model of a robot manipulator [17-19,25-26]. One of the difficulties in using microprocessors for the real-time control of manipulators lies in the complexity of computations required to compute a control law based on the dynamic model. We shall describe a cost-effective architecture, using currently available microprocessors, to compute in real-time the joint torques based on either Eq. (1) or Eq. (3) to servo control a manipulator. The design of this architecture is based on a functional decomposition of the equations of motion of a manipulator and is applicable to any manipulator with rotary/prismatic joints. This multiprocessor-based controller functions as a peripheral or an attached processor controller (APC) to some host computer which provides the desired trajectory set points. The APC interfaces the manipulator to the host system, monitors the manipulator motion, and performs control of the manipulator along a desired trajectory specified by the host system. In order to perform real-time control, the APC has to calculate the joint torques within the required sampling period. Our design goals were to analyze the EL and NE equations in order to decompose their computations into functional computational modules (CM's). Then based on these CM's, an appropriate computational structure using off-the-shelf microprocessors was designed to optimize its computational time. Architectures tuned to the functional flow of these two dynamic models are then examined to meet our design criterion (cost and speed criterion). This design approach yields a cost-effective architecture for real-time control of robot manipulators.

2.1. Computational Modules of Euler-Lagrange Equations of Motion

An examination of the EL equations of motion shows a certain amount of parallelism with a large amount of sequentialism in the flow of computation. The functional decomposition separates the inertial, Coriolis and centrifugal, and gravity terms of the EL equations of motion into a parallel form. Figure 1 shows the computational flow and data dependencies of the EL equations. All blocks represent computational modules, with the object being computed listed in the block, along with the number of computations required by that module. This figure shows that the gravitational, coupling (Coriolis and centrifugal), and inertial terms can be computed in parallel without any major dependencies on data from the other terms except for CM 5, which is used for computing both the coupling and inertial terms. However, the figure also shows that there is not much parallelism in computing within any of these major terms, but rather a highly sequential flow of computation. The main bottleneck in this algorithm is the computation of the coupling terms, especially in CM 4 and CM 7. These are highly sequential in nature; their computation depends on the data from the previous stage.

The flow of computation in the EL equations of motion suggests that the parallelism of the computation of the gravitational, coupling and inertial terms can be exploited by having multiple processing elements calculating these terms in parallel. The sequential nature of the algorithm can be exploited by a pipeline-like architecture which would have the CM's as stages in the pipeline.

The EL equations are not conducive to implementation using currently available hardware because the required number of computations is too high, even after taking the

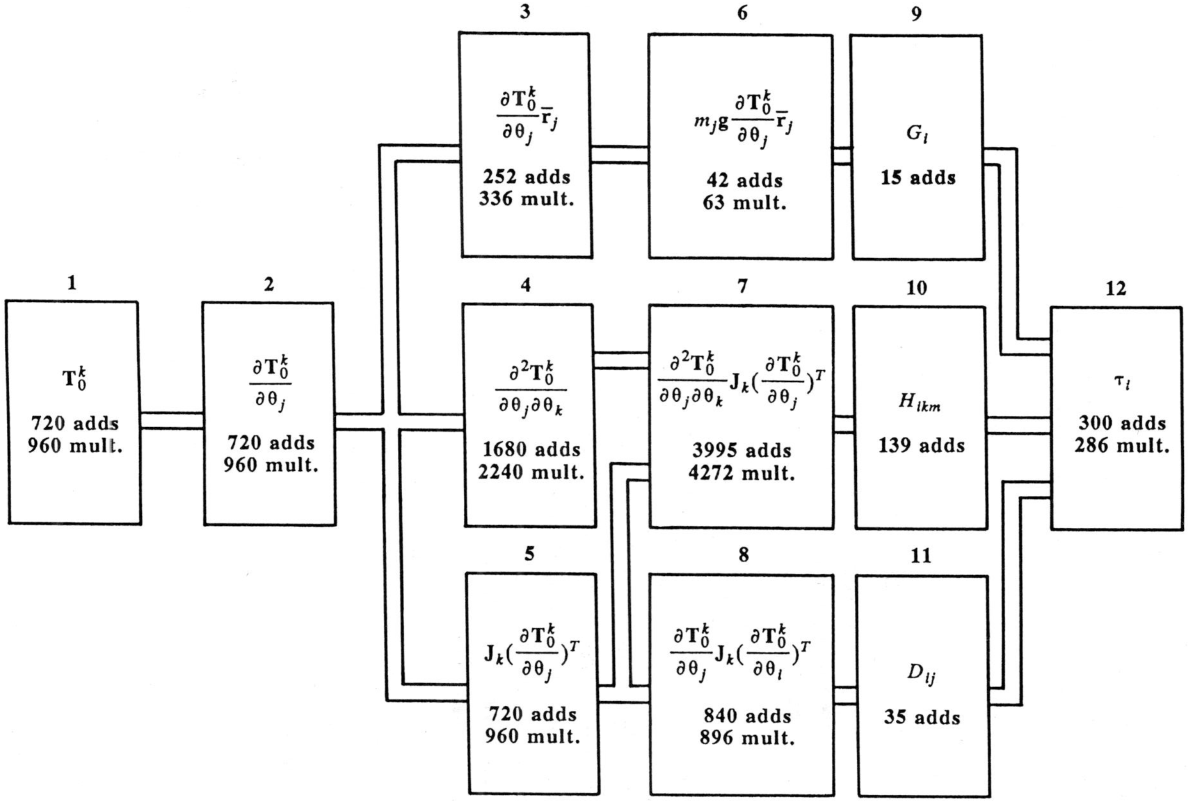

Figure 1. Computational Flow in EL Equations of Motion.

parallelism into account. Hollerbach [35] proposed a recursive EL formulation which increases the serialism in the algorithm. However, even this does not allow computation of the torques to be done in a timely fashion using currently available microprocessors, because the required number of computations is still very high. Targeting the architecture for controlling a particular robot could reduce the number of computations. However, our objective requires the proposed architecture to be able to handle any manipulator with rotary/prismatic joints. For these reasons, the EL equations were eliminated from consideration as the basis for the controller architecture.

2.2. Computational Modules of Newton-Euler Equations of Motion

The NE equations of motion are analyzed for their data dependencies and computational flow. The NE equations of motion lend themselves to being decomposed to the terms used to generate the recursive forward and backward equations. The flow of computation and data dependencies in these equations are shown in Figure 2. All CM's are again in blocks showing the object being computed and the amount of computation required for that module. This figure also indicates that these equations are highly serial in nature. The serial nature of the computational flow lends itself to a pipeline implementation [36]. The figure does not properly show the parallelism in this algorithm due to the recursion performed in the equations. Once a CM has computed a term for joint i, the CM can then start computing the same term for the next joint. Meanwhile, other CM's, requiring the results of this CM to perform their computation for joint i, can then start their computations. This allows multiple computational modules to be processing in parallel at any given time.

The bottleneck in the computation of the torques using the NE equations is the moment term (CM 3 and CM 7). However, as the figure shows, this term can be decomposed to two CM's, one (CM 3) can be calculated on the forward recursive pass, while the other (CM 7) is calculated on the backward recursive pass. This greatly increases the computational throughput.

The NE equations of motion are a good candidate for implementation due to their relatively low computation requirements and their recursive structure. The parallelism in the equations allows multiple processing elements, while the sequential nature of the equations lends itself to a pipeline architecture, both of which aid in increasing the computational throughput. An architecture based on this functional decomposition of the NE equations is presented in the next subsection.

2.3. Multiprocessor-Based Architecture

In our application, a global pipeline architecture for computing the joint torques from the NE equations is not appropriate because it can cause the manipulator to oscillate. This will occur if the sampling period is less than the time required to calculate the

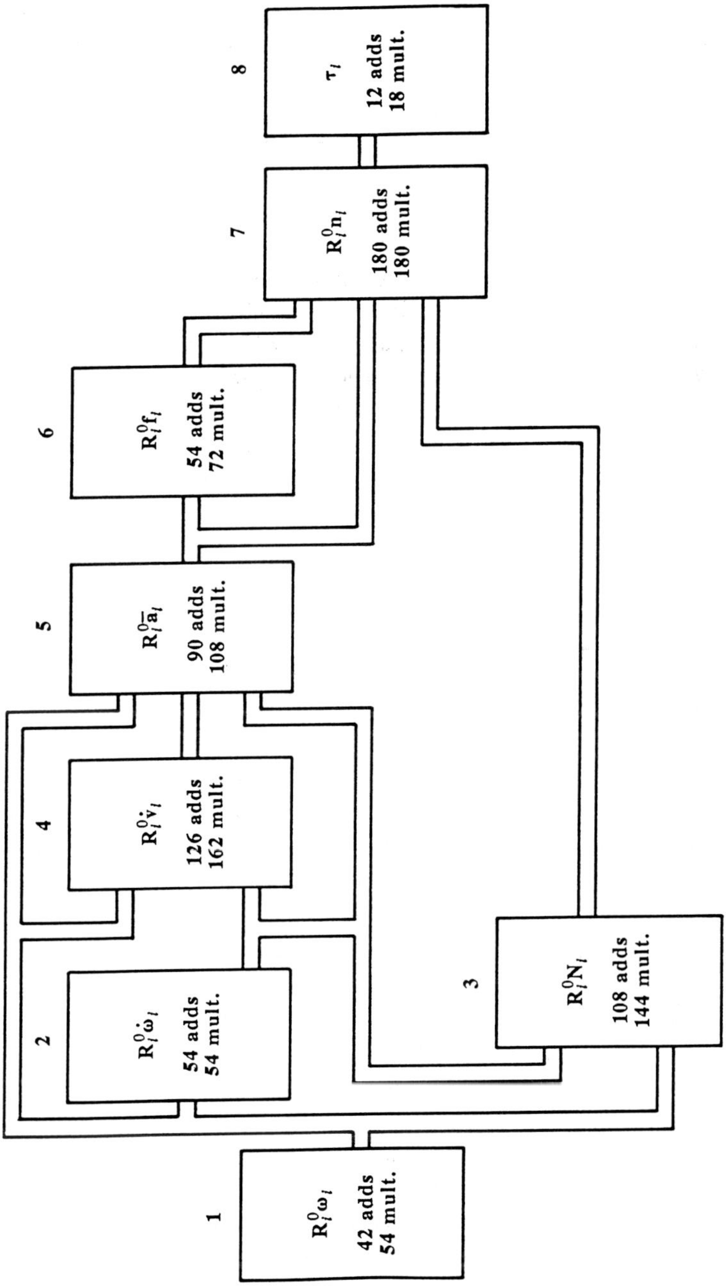

Figure 2. Computational Flow in NE Equations of Motion.

joint torques through the pipeline. That is, the sampling period is less than the initial delay time of the pipeline. In this case, the joint torques for time t are computed, and the appropriate control applied, while other joint torques for times $t+1$ through $t+m-1$ are still in the pipeline being computed, where m is the number of stages in the pipeline. These torques cannot take the correction torques into account, and hence will either overcompensate or undercompensate the manipulator. New joint torques being calculated will try to compensate for this "overdriving," thus causing oscillation. To eliminate this problem, pipelining is allowed only if the output of the pipeline takes less time to compute than the sampling period. This prevents acting on the system without taking an applied torque into account. However, the main benefit of a pipeline system is increased computational throughput, as subsequent calculations can be initiated before the result of the current calculation is produced. Pipelining can be utilized at a local level to calculate partial results. This allows overlapping of computations, leading to reduced computational delay.

The APC is required to calculate the joint torques, as in Eq. (3) and Appendix B, in 1.5 milliseconds from the time the joint variables $(q_i , \dot{q}_i)$ are obtained from the robot arm (via encoders, tachometers, etc.). The 1.5-millisecond period was determined by the criterion that the control would be updated at a rate of 20 times the natural frequency of the robot arm. In the case of the PUMA robot arm, the natural frequency was found experimentally to be approximately 15 Hz, which leads to a 300-Hz frequency for updating the control. This yields an approximate 3-millisecond sampling period. All joint samplings, conversions, scaling, coordinate transformations and control computations must be done in this period of time. We allocated 1.5 milliseconds for the controller calculations as in Eq. (1) (or Eq. (3)), and 1.5 milliseconds for all the other calculations. To appreciate the amount of computation required, this corresponds to 0.98 million arithmetic operations per second for the NE equations of motion (neglecting processing overhead, and assuming that multiplications and additions take approximately the same amount of time to perform, which is the case in many floating point systems).

The architecture that satisfies the above required computation rate is based on the NE equations, taking the parallel and sequential nature of the NE algorithm into account. A block diagram of the multiprocessor-based architecture is shown in Figure 3. As the diagram shows, there are multiple processing elements, each of which performs the task of calculating one (or more) CM's in the NE equations of motion. Each CM is allocated to a separate processor, except for CM's 1, 3, 7, and 8. CM 1 and CM 8 are allocated to the same processor, as they will never be calculated at the same time. This also applies to CM 3 and CM 7, as all ${}^i\mathbf{R}_0\mathbf{N}_i$ need to be computed before any ${}^i\mathbf{R}_0\mathbf{n}_i$ can be calculated. Figure 4 shows the timing of the computations in various processors. This figure shows that using six processors allows six CM's to be processing concurrently. There is a tradeoff between the number of processors used and the computational time for the control. The processors could be more fully utilized by reducing the number of

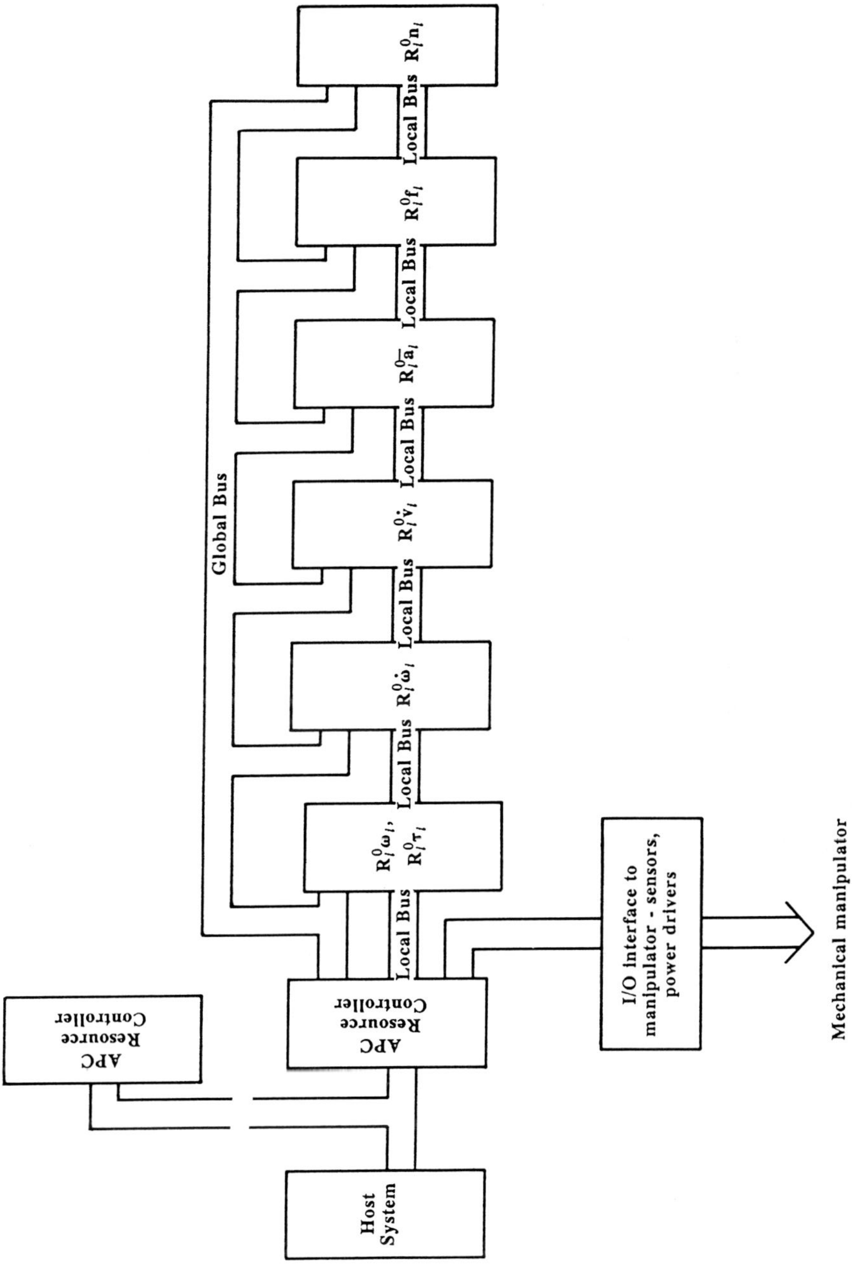

Figure 3. Proposed Architecture Based on NE Equations of Motion.

Stage \ Time	0–1	1–2	2–3	3–4	4–5	5–6	6–7	7–8	8–9	9–10	10–11	11–12	12–13	13–14	14–15
Process 1	${}^1R_0\omega_1$	${}^2R_0\omega_2$	${}^3R_0\omega_3$	${}^4R_0\omega_4$	${}^5R_0\omega_5$	${}^6R_0\omega_6$					τ_6	τ_5	τ_4	τ_3	τ_2	τ_1
Process 2	${}^1R_0\dot{\omega}_1$	${}^2R_0\dot{\omega}_2$	${}^3R_0\dot{\omega}_3$	${}^4R_0\dot{\omega}_4$	${}^5R_0\dot{\omega}_5$	${}^6R_0\dot{\omega}_6$									
Process 3		${}^1R_0\dot{v}_1$	${}^2R_0\dot{v}_2$	${}^3R_0\dot{v}_3$	${}^4R_0\dot{v}_4$	${}^5R_0\dot{v}_5$	${}^6R_0\dot{v}_6$								
Process 4		${}^1R_0\bar{a}_1$ terms dependent on ω_1, $\dot{\omega}_1$	${}^1R_0\bar{a}_1$	${}^2R_0\bar{a}_2$	${}^3R_0\bar{a}_3$	${}^4R_0\bar{a}_4$	${}^5R_0\bar{a}_5$	${}^6R_0\bar{a}_6$							
Process 5				${}^1R_0f_1$, ${}^1R_0n_1$ terms dependent on $\bar{a}_1$	${}^2R_0f_2$, ${}^2R_0n_2$ terms dependent on $\bar{a}_2$	${}^3R_0f_3$, ${}^3R_0n_3$ terms dependent on $\bar{a}_3$	${}^4R_0f_4$, ${}^4R_0n_4$ terms dependent on $\bar{a}_4$	${}^5R_0f_5$, ${}^5R_0n_5$ terms dependent on $\bar{a}_5$	${}^6R_0f_6$	${}^5R_0f_5$	${}^4R_0f_4$	${}^3R_0f_3$	${}^2R_0f_2$	${}^1R_0f_1$		
Process 6		${}^1R_0n_1$ terms dependent on ω_1, $\dot{\omega}_1$	${}^2R_0n_2$ terms dependent on ω_2, $\dot{\omega}_2$	${}^3R_0n_3$ terms dependent on ω_3, $\dot{\omega}_3$	${}^4R_0n_4$ terms dependent on ω_4, $\dot{\omega}_4$	${}^5R_0n_5$ terms dependent on ω_5, $\dot{\omega}_5$	${}^6R_0n_6$ terms dependent on ω_6, $\dot{\omega}_6$		${}^6R_0n_6$ terms dependent on $\bar{a}_6$	${}^6R_0n_6$	${}^5R_0n_5$	${}^4R_0n_4$	${}^3R_0n_3$	${}^2R_0n_2$	${}^1R_0n_1$	

Figure 4. Timing Diagram of Proposed Architecture.

processors, although the processing time for the control would increase. Alternatively, processing time for the control could be decreased by increasing the number of processors. In our case, any architecture that can meet the 1.5-millisecond computational time constraint will suffice. We chose six processors for this architecture because this allows a logical assignment of processors to CM's while providing a high degree of concurrency in computation. A lower number of processors would increase processing concurrency with a higher processor utilization. However, the mapping of the CM's to the processors would not be as straightforward, and the computational time would be increased. This mapping of a lower number of processors would be highly dependent on the number of joints. A change in the number of joints would require a different mapping if the computational load became too heavy for any processor. With six processors, the computational modules have been logically assigned to a processor with no dependence on the number of joints. The computational efficiency of this architecture tends to increase as the number of joints increases, since more terms can be processed using the pipeline structure of the architecture. With this architecture, processor utilization can be increased with few problems; processors which complete their computations in less than the required time can be used to assist other processors in their computation or assist the supervisor in other system tasks, such as transformation, scaling, and conversions. This reconfiguration can be done in software and may require further investigation for efficient reconfiguration.

This architecture takes advantage of the sequential nature of the NE equations of motion via software pipelining. The processing elements compute a term for link i, then broadcast it to all other processing elements that require this data for computation of their terms for link i. Once this broadcast occurs, if the receiving processor is free, it can then immediately start computing its term; otherwise that processor can start the computation once it is free. A processor that has finished computing a term for link i can then start computing the same term for link $i+1$ (or $i-1$). For example, completion of ${}^i\mathbf{R}_0\dot{\mathbf{v}}_i$ will allow ${}^i\mathbf{R}_0\dot{\mathbf{v}}_{i+1}$ to be computed, while completion of ${}^i\mathbf{R}_0\mathbf{n}_i$ will allow ${}^i\mathbf{R}_0\mathbf{n}_{i-1}$ to be computed. Certain terms in the force and moment computations are dependent on results calculated in the forward recursive equations and can be computed on the forward pass, leaving only the propagation of the force and moment terms to be computed in the backward pass. This reduces the number of computations per pass from 678 multiplications and 597 additions, as seen by a sequential processor, to 351 multiplications and 306 additions for a six-link PUMA robot. The torque terms are generated sequentially, with the torque for link 6 generated after 251 multiplications and 207 additions are performed, while the torque for link 1 is generated after all 351 multiplications and 306 additions are performed. This contrasts to 597 additions and 678 multiplications for the torque calculations if done sequentially. This corresponds to an approximately 49% reduction in the computational time for the joint torques when using a Motorola MC68020 microprocessor.

This architecture also provides an additional seventh processor which is used as a system supervisor. This processor handles all data transfer to a host system, as well as interpreting the action required by the host. This processor also handles all the I/O associated with the manipulator being controlled (e.g., sampling sensors and power drivers for the joint motors). The supervisor communicates to a host via a high speed synchronous serial link, thus allowing virtually any host to be connected to it. This serial link could also be used to connect multiple manipulators to the same host, as well as to allow one APC to communicate with another APC. The supervisor communicates with other processors in the APC via a global bus, while it also has a local bus to the processor that calculates the torque and angular velocity terms. This is done because the supervisor provides the data required to calculate the angular velocity from the sensors, while the supervisor also needs the torque terms to do the control. The supervisor processor could also handle any user interface (teach pendant, terminal, etc.) if needed. The supervisor processor could even handle the host's function of controlling the manipulator, though the throughput would be degraded. If required, an additional processor could be put on the global bus to perform this function.

The interconnection between processors is provided in two ways, one for local processor communication and the other for global processor communication. Any processor which is dependent on data from another processor is defined to be a local processor to that other processor. This interconnection scheme allows any processor to communicate with any other processor, while also allowing time critical transfers to proceed without having to arbitrate for the global bus. This architecture would need to support a data rate of 0.5 Mbyte per second, including overhead for bus arbitration. This data rate was obtained by observing that the data being sent on the bus are vectors generated from the processing elements. For $n = 6$, there are 6 vectors broadcast per processing element with 6 processing elements in the system. The broadcast occurs once per calculation. Assuming 32 bit words, this yields a data rate of approximately 0.3 Mbyte per second. The data rate of 0.5 Mbyte per second is determined from this with system communication factored in. Most microprocessor buses currently in use support data rates around 5 MHz with data paths 16-32 bits wide; thus this data rate should not create any problem. The global bus would be one of the current popular buses (e.g., VMEbus and Multibus-II); the local bus would be a fairly simple bus, with not much overhead required for accessing it.

This multiple processor arrangement also has the advantage of enhanced reliability, as failure of one processor can be gracefully handled with software reconfiguration. The manipulator control can still be accomplished with a slight degradation in computational time. System fault tolerance is an item for future research. This architecture also allows processors that have completed their computations for a particular pass to help out with computations for other terms or with some other tasks.

Although faster microprocessors are being developed at a rapid pace, this architecture can be implemented and meet our design criterion using the less expensive Intel 80286 or the MC68020 microprocessor using integer arithmetic. The use of floating point arithmetic with these processors would still take too much time to compute the joint torques. The lack of precision of integer arithmetic can be overcome by using scaling. The MC68020 microprocessor provides more precision with its 32 bit arithmetic so it is preferable to the 80286. Implementation of this architecture using six 80286s would require 1.2 milliseconds for the control computations, while a MC68020 implementation would require 0.6 milliseconds for the control computations. This is primarily due to the higher rated operating speed of the MC68020. The cost of this architecture is estimated to be 2,000 dollars, which is very competitive and very flexible for interfacing to any host, as well as to other APC's. Several architectures which satisfy our performance and/or cost criteria are also examined as alternatives to the multiprocessor-based architecture. Details can be found in [25].

3. EFFICIENT PARALLEL ALGORITHMS

Last section we described a cost-effective multiprocessor-based architecture for real-time control of robot manipulators based on a functional decomposition of the NE equations of motion. The architecture was based on the recursive structure of the NE equations of motion which are conducive to a pipeline implementation. Although the multiprocessor-based pipelined controller was able to reduce the number of computations to 351 multiplications and 306 additions for a six-link PUMA robot as compared to 678 multiplications and 597 additions for the torque calculations if done sequentially, the computational complexity of the NE equations of motion remains to be of order $O(n)$. To reduce the order of computation, further improvement in the computational efficiency of the NE equations of motion could be achieved by reformulating the equations of motion to be computed on single-instruction-stream multiple-data-stream (SIMD) computers.

The recursive structure of the NE equations of motion is obviously well suited to be computed on single-instruction-stream single-data-stream (SISD) computers. It is, however, not an efficient parallel processing for SIMD computers that are capable of performing many simultaneous operations. The NE equations of motion can be reformulated in a homogeneous linear recurrence form, and the *recursive doubling* algorithm [37-38] for parallel solution of linear recurrence problems can be utilized to compute the joint torques in SIMD computers. This section addresses 1) the limitation of speeding up the computation of the inverse dynamics problem, based on the NE equations of motion, of a manipulator running on p processors, where $1 \le p \le n$; this leads to the derivation of the time lower bound of computing the inverse dynamics of an n-link robot manipulator parallelly using p processors; 2) the reformulation of the NE equations of motion in a homogeneous linear recurrence form, and the use of the recursive doubling algorithm to achieve the time lower bound computation; 3) the development of an efficient p-fold parallel algorithm to overcome the recurrence problem of the NE equations of motion to achieve the time lower bound computation. The p-fold parallel algorithm can be best described as consisting of p-parallel blocks with pipelined elements within each parallel block. Taking advantage of the flexible properties of an SIMD machine, the algorithm can be easily implemented using a group of general-purpose microprocessors. A modified inverse perfect shuffle interconnection scheme can be efficiently used to interconnect the p processors. Finally, the parallel algorithm is also susceptible to a systolic pipelined architecture implementation. The most significance of this reformulation of the NE equations of motion for parallel machines is the reduction of computations from the time order of linearly proportional to the number of degrees of freedom, $O(n)$, to the time order of logarithmic proportional to the number of degrees of freedom, $O(\log_2 n)$.

3.1. Time Lower Bound Of Inverse Dynamics Problem

Since we are utilizing p processors, where $1 \leq p \leq n$, to compute the joint torques, we need to address the limitation of speeding up the computation based on the NE equations of motion of a manipulator. The purpose is to establish a time lower bound for the inverse dynamics computation problem so that several efficient computational schemes can be compared and contrasted. Let $T(n)$ be a function indicating the time lower bound of an algorithm which computes the inverse dynamics problem. If we have another algorithm whose computing time is of the same order as $T(n)$, then we know that asymptotically we can do no better with this algorithm.

Deriving a time lower bound on the inverse dynamics problem is perhaps more difficult than devising efficient algorithms for it. This is because a time lower bound indicates the time limitation of all possible algorithms for solving the given problem. Since we cannot enumerate and analyze all possible algorithms, time lower bound proofs are often harder to obtain.

In deriving the time lower bound of evaluating the inverse dynamics, the inverse dynamics problem and a linear arithmetic expression must be defined, then the time lower bound of evaluating a linear arithmetic expression using p processors will be extended to prove the time lower bound theorem for computing the joint torques of an n-link manipulator using p processors.

Definition 1: Inverse dynamic problem. Given the joint positions and velocities $\{q_j(t), \dot{q}_j(t)\}_{j=1}^{n}$, which describe the state of the manipulator at time t, expressed in the base coordinate system, together with the desired joint accelerations $\{\ddot{q}_j(t)\}_{j=1}^{n}$, solve the dynamic equations of motion for the joint torques $\{\tau_j(t)\}_{j=1}^{n}$ as follows:

$$\boldsymbol{\tau}(t) = \mathbf{f}(\mathbf{q}(t), \dot{\mathbf{q}}(t), \ddot{\mathbf{q}}(t)) \tag{4}$$

where $\boldsymbol{\tau}(t) = (\tau_1, \tau_2, \cdots, \tau_n)^T$, $\mathbf{q}(t) = (q_1, q_2, \cdots, q_n)^T$, $\dot{\mathbf{q}}(t) = (\dot{q}_1, \dot{q}_2, \cdots, \dot{q}_n)^T$, and $\ddot{\mathbf{q}}(t) = (\ddot{q}_1, \ddot{q}_2, \cdots, \ddot{q}_n)^T$.

Definition 2: A linear arithmetic expression is any well-formed string composed of four arithmetic operators $(+, -, \times, /)$ or, for convenience, two operators + (or −), × (or /), left and right parentheses, and atoms, which are constants or variables. We denote a linear arithmetic expression E of n distinct atoms by $E\langle n \rangle$; for example, $E\langle 4 \rangle : a + b - c / d$.

Let us denote $T_p\, [f_1(\cdot), f_2(\cdot), \cdots, f_n(\cdot)]$ be the *minimum* computing time needed to evaluate $[f_1(\cdot), f_2(\cdot), \cdots, f_n(\cdot)]$ running on p processors, then the time lower bound of evaluating a linear arithmetic expression $E\langle n \rangle$ using p processors, $T_p\, [E\langle n \rangle]$, can be stated in Lemma 1.

Lemma 1: The time lower bound of $T_p\, [E\langle n \rangle]$ [39], the shortest parallel time to evaluate a linear arithmetic expression $E\langle n \rangle$ using p processors, is bounded below by and equal to $O(\lceil n/p \rceil + \lceil \log_2 p \rceil)$; that is,

$$T_p\, [E\langle n \rangle] \geq O(\lceil n/p \rceil + \lceil \log_2 p \rceil)\,.$$

The proof of Lemma 1 can be found in [39]. Based on the above definitions, notation, and lemma, we are ready to state and prove the time lower bound theorem for computing the joint torques of an n-link manipulator using p processors.

Theorem 1: The shortest parallel time to evaluate the joint torques $\{\tau_j(t)\}_{j=1}^n$ of an n-link manipulator from the Newton-Euler equations of motion using p processors, where $1 \leq p \leq n$, is bounded below by $O(k_1 \lceil n/p \rceil + k_2 \lceil \log_2 p \rceil)$, where k_1 and k_2 are specified constants.

Proof: Using the Newton-Euler equations of motion, the inverse dynamics problem can be considered as computing a set of arithmetic expressions (not necessarily linear) which result in obtaining the joint torques. Each joint torque $\tau_j(t)$ of joint j can be expressed as an arithmetic expression containing at least $3n$ atoms: n joint positions $\{q_j(t)\}_{j=1}^n$, n joint velocities $\{\dot{q}_j(t)\}_{j=1}^n$, and n joint accelerations $\{\ddot{q}_j(t)\}_{j=1}^n$; that is,

$$\begin{aligned}\tau_j(t) &= f_j\,(q_1, q_2, \cdots, q_n, \dot{q}_1, \dot{q}_2, \cdots, \dot{q}_n, \ddot{q}_1, \ddot{q}_2, \cdots, \ddot{q}_n) \\ &= f_j\,(\mathbf{q}(t), \dot{\mathbf{q}}(t), \ddot{\mathbf{q}}(t))\end{aligned}$$

where $f_j(\cdot)$ is an arithmetic expression and *not* necessary linear. It can be shown that the following inequalities are true:

$$\begin{aligned}T_p\,[\tau_1, \tau_2, \cdots, \tau_n] &\geq T_p[\text{To evaluate a specified joint torque } \tau_j,\ 1 \leq j \leq n] \qquad (5) \\ &= T_p\,[f_j(\mathbf{q}, \dot{\mathbf{q}}, \ddot{\mathbf{q}})].\end{aligned}$$

Using Lemma 1, we have

$$T_p\,[f_j(\mathbf{q}\,,\dot{\mathbf{q}}\,,\ddot{\mathbf{q}})] \geq T_p\,[E<3n>] \geq O(\lceil 3n/p\rceil + \lceil \log_2 p\rceil)\,. \tag{6}$$

Combining Eq. (5) and Eq. (6), we have

$$T_p\,[\tau_1, \tau_2, \cdots, \tau_n] \geq O(\lceil 3n/p\rceil + \lceil \log_2 p\rceil) = O(c(\lceil 3n/p\rceil + \lceil \log_2 p\rceil)) \tag{7}$$

where c is a constant. We can rewrite Eq. (7) in a more general form without changing the order of the lower bound

$$T_p\,[\tau_1, \tau_2, \cdots, \tau_n] \geq O(k_1\lceil n/p\rceil + k_2\lceil \log_2 p\rceil) \tag{8}$$

where k_1 and k_2 are two specified constants. ■

Two extreme cases can be observed from Theorem 1:

(a) If $p = 1$, then the shortest computing time $T_p\,[\tau_1, \tau_2, \cdots, \tau_n]$ is not lower than $O(n)$.

(b) If $p = n$, then the shortest parallel computing time $T_p\,[\tau_1, \tau_2, \cdots, \tau_n]$ is not lower than $O(\lceil \log_2 n\rceil)$.

Theorem 1 indicates that the shortest parallel computing time $T_p\,[\tau_1, \cdots, \tau_n]$ for the inverse dynamics problem is bounded below by and not necessary equal to $O(k_1\lceil n/p\rceil + k_2\lceil \log_2 p\rceil)$. In other words, an efficient algorithm with p-fold parallelism may not have a computing time of the same order as $O(k_1\lceil n/p\rceil + k_2\lceil \log_2 p\rceil)$. However, if a parallel algorithm possesses the time lower bound, then it must be the most efficient algorithm of evaluating the inverse dynamics. Since the NE formulation possesses a time order of $O(n)$ as shown in Appendix B, it is the most efficient algorithm of evaluating the inverse dynamics running in uniprocessor computers. However, Theorem 1 also indicates that a better solution is to find an efficient parallel algorithm, running on p processors, $1 \leq p \leq n$, that possesses a time complexity order of $O(k_1\lceil n/p\rceil + k_2\lceil \log_2 p\rceil)$. To achieve this time order, the NE equations of motion must be reformulated in a homogeneous linear recurrence form, and the recursive doubling algorithm for parallel solution of linear recurrence problems can be utilized to compute the joint torques in SIMD computers. The linear recurrence problem and the recursive doubling algorithm are discussed in the next subsection, then the inverse dynamics computation is formulated as a linear recurrence problem and the recursive doubling algorithm is used to solve the computations in the time order of $O(\lceil \log_2 n\rceil)$. The parallel algorithm utilizing the recursive doubling technique for computing the inverse dynamics is discussed in subsection 3.2. An efficient p-fold parallel algorithm that achieves the time lower bound is discussed in subsection 3.4. Finally, an efficient interconnection scheme for the p processors for the recursive doubling algorithm is discussed in subsection 3.5, and the VLSI implementation of the parallel algorithm is discussed in subsection 3.6.

3.2. Linear Recurrence Problem and Recursive Doubling Algorithm

The recursive Newton-Euler equations of motion are very efficient in evaluating the inverse dynamics whether they are formulated in the base coordinate frame (see Table 1 on the next page) or in the link coordinate frames (see Appendix B). The clear advantage of referencing both the dynamics and kinematics to the link coordinates is to obviate a great deal of coordinate transformations and to allow the inertia tensor to be constant in each link coordinate frame, which results in a much faster computation in a uniprocessor computer. However, the recursive structure of this formulation is in an inhomogeneous linear recursive form; that is, $\boldsymbol{\omega}_i = a_i \boldsymbol{\omega}_{i-1} + b_i$, where $a_i \equiv {}^i\mathbf{R}_{i-1}$ and $b_i \equiv {}^i\mathbf{R}_{i-1}\, \mathbf{z}_0 \dot{q}_i$†. When parallel computation is considered, this inhomogeneous linear recursive form is less attractive and requires more calculations and arrangements for parallel processing than the homogeneous linear recursive form. The NE equations of motion formulated in the base coordinates can be re-arranged and transformed into a homogeneous linear recurrence form; that is, $\boldsymbol{\omega}_i = \boldsymbol{\omega}_{i-1} + \dot{q}_i\, \mathbf{z}_{i-1}$, which is more suitable for parallel processing on an SIMD computer, yielding a much shorter computing time.

Kogge and Stone [37] developed an efficient technique, called recursive doubling, to solve a large class of recurrence problems on parallel (SIMD) computers. An mth-order recurrence problem is defined as the computation of a sequence of n elements denoted as $x(1), x(2), \cdots, x(n)$, given only a set of initial conditions $(x(0), \cdots, x(-m+1))$ and a set of equations relating each $x(i)$ to m previous x's elements of the sequence, $x(i) = f_i(\mathbf{c}(i), x(i-1), \cdots, x(i-m))$, and $\mathbf{c}(i)$ is some vector of parameters. An example of an mth-order recurrence problem is the description of a discrete-time linear system in which the state of the system at time i is a linear function of the state at time $i-1$: Given $x(0)$, the state $x(i)$ can be recursively computed from these equations:

$$x(1) = a(1)x(0) + b(1)$$

$$\vdots$$

$$x(i) = a(i)x(i-1) + b(i)$$

$$\vdots$$

$$x(n) = a(n)x(n-1) + b(n)$$

where $a(i)$ and $b(i)$ represent the known internal dynamics of the linear system. On a serial (SISD) computer, we first use the initial conditions to compute one new $x(i)$, then using the new $x(i)$ to compute $x(i+1)$, and so on until all n $x(i)$'s elements are computed. Hence, the computation time is proportional to n, and solving a linear recursive equation on a uniprocessor computer has a time complexity of $O(n)$. This time

† a_i and b_i are used here as either matrices or vectors which must be computed in advance.

Table 1. Recursive Newton-Euler Equations of Motion Referenced to the Base Coordinate System.

Forward (or Outward) Equations: $i = 1, 2, \cdots, n$

$$\boldsymbol{\omega}_i = \boldsymbol{\omega}_{i-1} + \dot{q}_i \mathbf{z}_{i-1} (1 - \lambda_i)$$

$$\dot{\boldsymbol{\omega}}_i = \dot{\boldsymbol{\omega}}_{i-1} + (\ddot{q}_i \mathbf{z}_{i-1} + \boldsymbol{\omega}_{i-1} \times \dot{q}_i \mathbf{z}_{i-1})(1 - \lambda_i)$$

$$\dot{\mathbf{v}}_i = \dot{\boldsymbol{\omega}}_i \times \mathbf{p}_i^* + \boldsymbol{\omega}_i \times (\boldsymbol{\omega}_i \times \mathbf{p}_i^*) + \dot{\mathbf{v}}_{i-1} + (2\boldsymbol{\omega}_i \times \dot{q}_i \mathbf{z}_{i-1} + \ddot{q}_i \mathbf{z}_{i-1})\lambda_i$$

$$\overline{\mathbf{a}}_i = \dot{\boldsymbol{\omega}}_i \times \overline{\mathbf{s}}_i + \boldsymbol{\omega}_i \times (\boldsymbol{\omega}_i \times \overline{\mathbf{s}}_i) + \dot{\mathbf{v}}_i$$

Backward (or Inward) Equations: $i = n, n-1, \cdots, 1$

$$\mathbf{f}_i = m_i \overline{\mathbf{a}}_i + \mathbf{f}_{i+1}$$

$$\mathbf{n}_i = \mathbf{n}_{i+1} + \mathbf{p}_i^* \times \mathbf{f}_{i+1} + (\mathbf{p}_i^* + \overline{\mathbf{s}}_i) \times m_i \overline{\mathbf{a}}_i + \mathbf{I}_i \dot{\boldsymbol{\omega}}_i + \boldsymbol{\omega}_i \times (\mathbf{I}_i \boldsymbol{\omega}_i)$$

$$\tau_i = (\mathbf{n}_i^T \mathbf{z}_{i-1})(1 - \lambda_i) + (\mathbf{f}_i^T \mathbf{z}_{i-1})\lambda_i + b_i \dot{q}_i$$

where b_i is the viscous damping coefficient for joint i, $\lambda_i = 0$ if joint i is rotary, and $\lambda_i = 1$ if joint i is prismatic. The "usual" initial conditions are: $\boldsymbol{\omega}_0 = \dot{\boldsymbol{\omega}}_0 = \mathbf{v}_0 = \mathbf{0}$ and $\dot{\mathbf{v}}_0 = (g_x, g_y, g_z)^T$ (to include gravity), where $|\mathbf{g}| = 9.8062 \, m/s^2$. The kinematic variables used here are defined in Appendix B.

complexity can be reduced to logarithmic proportional to the number of degrees of freedom of the manipulator if a parallel solution, called recursive doubling, is applied to solve the linear recurrence problems in an n-processor SIMD computer. Recursive doubling involves the splitting of the computation of such problem into two *equally* complex subproblems. The evaluation of these subproblems can be performed simultaneously in two separate processors. By repeating the same procedure, each subproblem can be further split and spread over more processors. For the computation of a sequence of n elements, there will be $(n+1)/2^k$ parallel operations at the kth splitting until $\lceil \log_2(n+1) \rceil$ splittings. The resulting algorithm computes the entire series $x(0), x(1), \cdots, x(n)$ in time complexity order of $O(\lceil \log_2(n+1) \rceil)$ on an n-processor computer. Thus, the recursive doubling algorithm achieves the time lower bound and hence is the best solution for the linear recurrence problem.

For the manipulator dynamics problem (forward and inverse dynamics problems), we are only interested in the simplest class of recurrence problem — the first-order linear recurrence problem in which $x(i)$ depends only on $x(i-1)$. In general, the first-order linear recurrence problem can be stated as: Given $a(j) \neq 0$ and $b(j)$, $0 \leq j \leq n$, and the recursive equation

$$x(i) = a(i) * x(i-1) + b(i), \tag{9}$$

where $*$ and $+$ may be scalar (or matrix) multiplication and scalar (or vector or matrix) addition, respectively, find $x(1)$, $x(2)$, $\cdots$,$x(n)$ by a parallel algorithm running on an n-processor computer. If either $a(i)$'s are identities or $b(i)$'s are null for all i, then this becomes the first-order *homogeneous* linear recurrence problem. Otherwise, it becomes the first-order *inhomogeneous* linear recurrence problem.

For the inverse dynamics problem, we are only interested in the homogeneous linear recurrence problem of size $(n+1)$ and can be described as follows: Given $a(i)$, $0 \leq i \leq n$, and $x(0) = a(0) \neq$ identity, find $x(1)$, $x(2)$, $\cdots$, $x(n)$ by a parallel algorithm running on an SIMD computer with n processors, where

$$x(i) = x(i-1) * a(i) \tag{10}$$

and "$*$" denotes an associative operator, which can be a matrix product or addition, a vector dot product or addition, etc. If $a(0)$ is equal to an identity, then the linear recurrence of size $(n+1)$ can be reduced to the case of size n by taking a shift operation as follows:

$$a(0) \leftarrow a(1),\ a(1) \leftarrow a(2),\ \cdots,\ a(i-1) \leftarrow a(i),\ \cdots,\ a(n-1) \leftarrow a(n),$$

where the back-arrow "$\leftarrow$" denotes a replacing operation. The recursive doubling algorithm solves the homogeneous linear recurrence form in Eq. (10) by splitting the computations of a serial associative operation, for example, $x(3) = a(3) * a(2) * a(1) * a(0)$

into two suboperations as $x(3) = [a(3) * a(2)] * [a(1) * a(0)]$ whose evaluations can be performed simultaneously in two processors during the first splitting, one computing $[a(3) * a(2)]$ and the other computing $[a(1) * a(0)]$. For an arbitrary $x(n)$, the generalization of the idea allows $(n+1)/2$ parallel operations at the first splitting, $(n+1)/4$ at the second, and $(n+1)/2^k$ at the kth until $\lceil \log_2(n+1) \rceil$ splittings, then $x(n)$ is computed with one final operation. Similarly, it can be shown that $x(i)$ can be computed in $\lceil \log_2(i+1) \rceil$ splittings, where $1 \le i \le n$. In other words, with the recursive doubling algorithm, $x(1), x(2), \cdots, x(n)$ can be computed concurrently no later than the time step $\lceil \log_2(n+1) \rceil$. Hence, the recursive doubling algorithm solves a homogeneous linear recurrence problem of size $(n+1)$ on an n-processor computer in time complexity order of $O(\lceil \log_2(n+1) \rceil)$.

■ **Example 1:** Let us give a simple example to illustrate the essence of recursive doubling in calculating linear recurrence equations. Suppose we are to find $x(n) = a(0) + a(1) + \cdots + a(n)$. For an SISD computer, it will take n successive additions for a time complexity of $O(n)$. We can rewrite the problem in the form of linear recurrence equations:

$$x(0) = a(0)$$

$$x(1) = x(0) + a(1)$$

$$x(2) = x(1) + a(2) \equiv a(0) + a(1) + a(2)$$

$$\cdots$$

$$x(i) = x(i-1) + a(i) \equiv a(0) + a(1) + \cdots + a(i)$$

$$\cdots$$

$$x(n) = x(n-1) + a(n) \equiv a(0) + a(1) + \cdots + a(n) .$$

If n is taken to be 7, the computation and splitting of $x(n)$ can be done as depicted in Figure 5. The oblique arrows show the passing of data from one processor to another. From the figure, we see that the computation is done in $\lceil \log_2(n+1) \rceil = 3$ additions, and at the same time, $x(0), x(1), \cdots, x(n)$ are all computed. ■

The recursive doubling algorithm is applied to evaluate the simultaneous solution of the entire sequence $x(1), x(2), \cdots, x(n)$ from Eq. (10) as depicted in Figure 6. This figure also shows that the intercommunication between the n processors of an SIMD computer is performed by executing an algorithm, called First Order Homogeneous Recurrence Algorithm (FOHRA), shown below as well as implemented on a cascade-structured network. This cascade-structured network provides interconnections, data fetching and accessing between the processors, and also indicates that the initial constant value $x(0) = a(0)$ can be stored in the memory of each processor, and avoids the complicated data transferring. The algorithm FOHRA is summarized below.

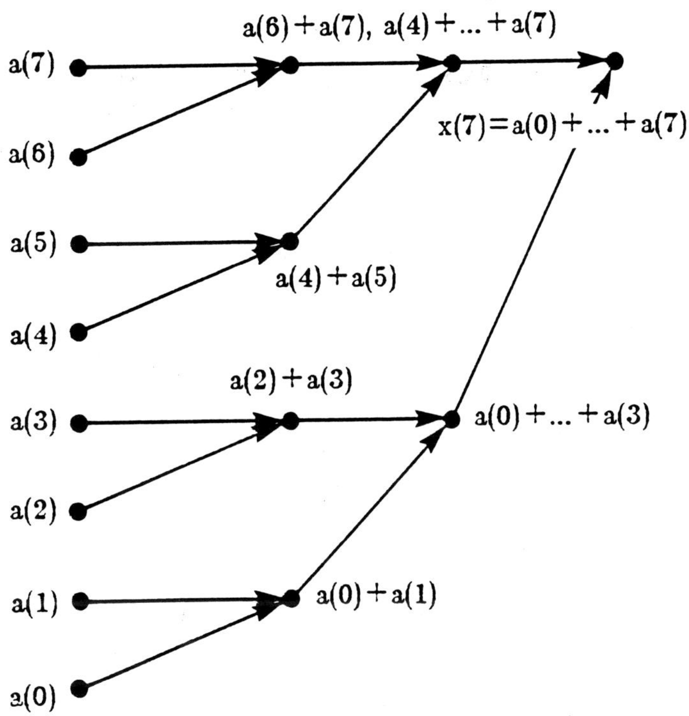

Figure 5. Parallel Computation of $x(7)$.

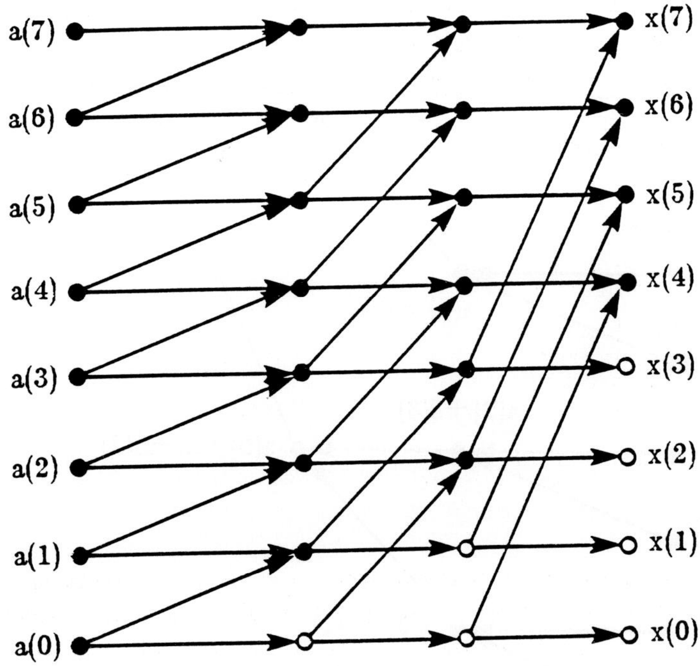

Figure 6.

Parallel Computation of $x(i) = x(i-1) * a(i)$.
Dark circles represent operations; open circles represent null operations.

Algorithm FOHRA (First Order Homogeneous Recurrence Algorithm). Given the terms $a(i)$, $0 \le i \le n$, this algorithm solves the homogeneous linear recurrence form in equation (10) by the recursive doubling algorithm.

F1. [Initialization.] Given the terms $a(i)$, $0 \le i \le n$, let $X^{(k)}(i)$ be the ith sequence at the kth splitting and $s = \lceil \log_2(n+1) \rceil$. Set the sequence at the initial step, $X^{(0)}(i) \leftarrow a(i)$, $0 \le i \le n$.

F2. [Compute $x(i)$ parallelly.]
for $k \leftarrow 1$ **to** s **do**

$$X^{(k)}(i) = \begin{cases} X^{(k-1)}(i-2^{k-1}) * X^{(k-1)}(i), & \text{if } 2^{k-1} \le i \le n \\ X^{(k-1)}(i), & \text{if } 0 \le i < 2^{k-1} \end{cases} \tag{11}$$

end_do

Set $x(i) \leftarrow X^{(s)}(i)$, $1 \le i \le n$.

END FOHRA

Similarly, the recursive doubling algorithm can be applied to evaluate the inhomogeneous linear recursive equation in Eq. (9) in obtaining the entire sequence $x(1), x(2), \cdots, x(n)$. The inhomogeneous linear recurrence problem can be solved by the First-Order Inhomogeneous Recurrence Algorithm (FOIHRA) which is stated here for convenience.

Algorithm FOIHRA (First-Order Inhomogeneous Recurrence Algorithm). Given the terms $a(i)$ and $b(i)$, $0 \le i \le n$, this algorithm computes the first-order inhomogeneous linear recurrence equation of the form in Eq. (9) by the recursive doubling algorithm.

I1. [Initialization.] Given the terms $a(i)$, $b(i)$, $0 \le i \le n$, let $X^{(k)}(i)$ and $Y^{(k)}(i)$ be the ith sequences at the kth level, and $s = \lceil \log_2(n+1) \rceil$. Set $X^{(0)}(i) \leftarrow a(i)$ and $Y^{(0)}(i) \leftarrow b(i)$ for $0 \le i \le n$.

I2. [Compute $x(i)$ parallelly.]
for $k \leftarrow 1$ **to** s **do**

$$X^{(k)}(i) = \begin{cases} X^{(k-1)}(i-2^{k-1}) * X^{(k-1)}(i), & \text{if } 2^{k-1} \le i \le n \\ X^{(k-1)}(i), & \text{if } 0 \le i < 2^{k-1} \end{cases} \tag{12}$$

$$Y^{(k)}(i) = \begin{cases} X^{(k-1)}(i) * Y^{(k-1)}(i-2^{k-1}) + Y^{(k-1)}(i), & \text{if } 2^{k-1} \le i \le n \\ Y^{(k-1)}(i), & \text{if } 0 \le i < 2^{k-1} \end{cases} \tag{13}$$

end_do

Set $x(i) \leftarrow Y^{(s)}(i)$, $1 \le i \le n$.

End FOIHRA

The "*" in steps F2 and I2 denotes an associative operator. Both FOHRA and FOIHRA algorithms have a computational complexity of $O(\lceil \log_2(n+1) \rceil)$ which is the time lower bound of the linear recurrence problem.

3.3. Inverse Dynamics Computation As A Linear Recurrence Problem

Let us apply the recursive doubling algorithm to compute the joint torques of an n-link manipulator based on the NE equations of motions and running on n processors. In order to apply the recursive doubling algorithm to the NE equations of motion, the equations of motion must be reformulated in the base coordinates (see Table 1) and re-arranged to a homogeneous linear recursive form. Prior to computing the joint torques from the NE equations of motion, some necessary input physical parameters relating the formulation must be given or evaluated in advance. These are as follows.

(a) The 3×3 rotation matrices ${}^{i-1}\mathbf{R}_i$, $i = 1, 2, \cdots, n$, which indicate the orientation of link i coordinates referenced to link $(i-1)$ coordinates, need to be evaluated in advance.

(b) ${}^{i}\mathbf{p}_i^*$ denotes the origin of link i coordinate frame from the origin of link $(i-1)$ coordinate frame, expressed with respect to link i coordinates. ${}^{i}\bar{\mathbf{s}}_i$ denotes the location of the center of mass of link i from the origin of link i coordinate frame, expressed with respect to link i coordinates. ${}^{i}\mathbf{I}_i$ denotes the inertia matrix of link i about its center of mass, expressed with respect to link i coordinates. Note that ${}^{i}\mathbf{p}_i^*$, ${}^{i}\bar{\mathbf{s}}_i$, and ${}^{i}\mathbf{I}_i$ must be given in advance and are constants when referred to their own link coordinates.

(c) λ_i is an indicator which specifies the motion of link i; $\lambda_i = 1$, when link i is rotational, and $\lambda_i = 0$, when link i is translational.

(d) Let $\boldsymbol{\omega}_0 = \dot{\boldsymbol{\omega}}_0 = \mathbf{0}$, $\dot{\mathbf{v}}_0 = (g_x, g_y, g_z)^T$ and $|\mathbf{g}| = 9.80621$ meters/sec^2. If non-zero external force $\mathbf{f}_e$ and external moment $\mathbf{n}_e$ are exerting on link n, then $\mathbf{f}_{n+1} = \mathbf{f}_e$ and $\mathbf{n}_{n+1} = \mathbf{n}_e$; otherwise, $\mathbf{f}_{n+1} = \mathbf{n}_{n+1} = \mathbf{0}$.

The procedure in computing the inverse dynamics from the NE equations of motion formulated in the base coordinates is to re-arrange the kinematic equations, which compute the angular velocity of each link, the angular and linear accelerations of each link, the linear acceleration of the center of mass of each link, and to re-arrange the dynamic equations, which relate the forces and moments exerted on successive links, in a homogeneous linear recursive form. The procedure of evaluating the NE equations of motion as a linear recurrence problem is given in the following parallel Newton-Euler computation procedure.

Step 1. Compute the 3×3 rotation matrices of link i coordinates with respect to the base coordinates, ${}^0\mathbf{R}_i$, $i = 1, 2, \cdots, n$,

$$ {}^0\mathbf{R}_i = {}^0\mathbf{R}_{i-1}\ {}^{i-1}\mathbf{R}_i \, . \tag{14} $$

Step 2. Compute $\mathbf{p}_i^*$, $\overline{\mathbf{s}}_i$, and $\mathbf{z}_i$ for $i = 1, 2, \cdots, n$,

$$ \mathbf{z}_i = {}^0\mathbf{R}_i\, \mathbf{z}_0 \tag{15} $$

$$ \mathbf{p}_i^* = {}^0\mathbf{R}_i\ {}^i\mathbf{p}_i^* \tag{16} $$

$$ \overline{\mathbf{s}}_i = {}^0\mathbf{R}_i\ {}^i\overline{\mathbf{s}}_i \tag{17} $$

where $\mathbf{z}_0 = (0, 0, 1)^T$. The evaluation of $\mathbf{z}_i$ only involves taking the third column of ${}^0\mathbf{R}_i$.

Step 3. Compute the angular velocity of link i

$$ b_i = \mathbf{z}_{i-1}\dot{q}_i\, \lambda_i \tag{18} $$

and

$$ \boldsymbol{\omega}_i = \boldsymbol{\omega}_{i-1} + b_i \, . \tag{19} $$

Step 4. Compute the angular acceleration of link i

$$ b_i = (\mathbf{z}_{i-1}\, \ddot{q}_i + \boldsymbol{\omega}_{i-1} \times \mathbf{z}_{i-1}\, \dot{q}_i)\, \lambda_i \tag{20} $$

and

$$ \dot{\boldsymbol{\omega}}_i = \dot{\boldsymbol{\omega}}_{i-1} + b_i \, . \tag{21} $$

Step 5. Compute the linear acceleration of link i

$$ b_i = \dot{\boldsymbol{\omega}}_i \times \mathbf{p}_i^* + \boldsymbol{\omega}_i \times (\boldsymbol{\omega}_i \times \mathbf{p}_i^*) + \tag{22} $$

$$ (\mathbf{z}_{i-1}\, \ddot{q}_i + 2\, \boldsymbol{\omega}_i \times (\mathbf{z}_{i-1}\, \dot{q}_i))\,(1 - \lambda_i) $$

and

$$ \dot{\mathbf{v}}_i = \dot{\mathbf{v}}_{i-1} + b_i \, . \tag{23} $$

Step 6. Compute the linear acceleration of the center of mass of link i

$$ \overline{\mathbf{a}}_i = \dot{\boldsymbol{\omega}}_i \times \overline{\mathbf{s}}_i + \boldsymbol{\omega}_i \times (\boldsymbol{\omega}_i \times \overline{\mathbf{s}}_i) + \dot{\mathbf{v}}_i \, . \tag{24} $$

Step 7. Compute the total external force exerted on link i at the center of mass

$$ \mathbf{F}_i = m_i \overline{\mathbf{a}}_i \, . \tag{25} $$

Step 8. Compute the total external moment exerted on link i at the center of mass

$$ \mathbf{N}_i = \mathbf{I}_i\, \dot{\boldsymbol{\omega}}_i + \boldsymbol{\omega}_i \times (\mathbf{I}_i\, \boldsymbol{\omega}_i) \, . \tag{26} $$

Since the calculations of evaluating $\mathbf{I}_i = {}^0\mathbf{R}_i\, {}^i\mathbf{I}_i\, {}^i\mathbf{R}_0$ were quite complicated,

Eq. (26) is modified to

$$^{i}\omega_i = {}^{i}\mathbf{R}_0\ \omega_i = ({}^{0}\mathbf{R}_i)^T\ \omega_i \tag{27}$$

$$^{i}\dot{\omega}_i = {}^{i}\mathbf{R}_0\ \dot{\omega}_i = ({}^{0}\mathbf{R}_i)^T\ \dot{\omega}_i \tag{28}$$

$$^{i}\mathbf{N}_i = {}^{i}\mathbf{I}_i\ {}^{i}\dot{\omega}_i + {}^{i}\omega_i \times ({}^{i}\mathbf{I}_i\ {}^{i}\omega_i) \tag{29}$$

$$\mathbf{N}_i = {}^{0}\mathbf{R}_i\ {}^{i}\mathbf{N}_i\ . \tag{30}$$

Step 9. Compute the force exerted on link i by link $(i-1)$

$$\mathbf{f}_i = \mathbf{f}_{i+1} + \mathbf{F}_i\ . \tag{31}$$

Step 10. Compute the moment exerted on link i by link $(i-1)$ at the

$$b_i = \mathbf{N}_i + (\mathbf{p}_i^* + \overline{\mathbf{s}}_i) \times \mathbf{F}_i + \mathbf{p}_i^* \times \mathbf{f}_{i+1} \tag{32}$$

and

$$\mathbf{n}_i = \mathbf{n}_{i+1} + b_i\ . \tag{33}$$

Step 11. Compute the joint torque at joint i

$$\tau_i = \begin{cases} (\mathbf{n}_i)^T\ \mathbf{z}_{i-1}\ , & \text{if } \lambda_i = 1 \\ (\mathbf{f}_i)^T\ \mathbf{z}_{i-1}\ , & \text{if } \lambda_i = 0\ . \end{cases} \tag{34}$$

All the above kinematic and dynamic variables are defined in Appendix B except they are expressed in the base coordinate system of the manipulator.

Equation (14) shows that the evaluation of ${}^{0}\mathbf{R}_i$ is a simple recursive equation in matrix product form. Equations (19), (21), (23), (31), and (33) only involve simple recursive vector addition form. The other equations in the NE equations of motion are not in recursive form and they can be computed parallelly. Both recursive matrix product form and recursive vector addition form are in the homogeneous linear recursive form, which is suitable for parallel evaluation based on the recursive doubling algorithm and capable of implementing the calculations on an SIMD machine.

By applying the recursive doubling algorithm to the above NE equations in the base coordinates for the PUMA robot arm in [1], the evaluation of the total computational complexity of the parallel Newton-Euler algorithm can be derived as follows:

(a) If $a(0) = {}^{0}\mathbf{R}_0 =$ identity matrix, then the parallel evaluation of Eq. (14) becomes a case of linear recurrence of size n, yielding $(27\lceil \log_2 n \rceil)$ scalar multiplications † and $(18\lceil \log_2 n \rceil)$ scalar additions †.

† Here we assume all the 9 elements of ${}^{i-1}\mathbf{R}_i$ are non-zero elements for ease of programming the computations. Since the (3,1) element of ${}^{i-1}\mathbf{R}_i$ is always zero, the numbers of scalar multiplications and additions can be taken to be $27\lceil \log_2 n \rceil - 6$ and $18\lceil \log_2 n \rceil - 2$, respectively.

(b) Equations (19), (21), (23), (31), and (33) all have the same recursive vector addition form, the parallel evaluation of Eq. (19) or Eq. (21) requires $3\lceil \log_2 n \rceil$ scalar additions for the case of $a(0) = \mathbf{0}$ (i.e., $a(0) = \boldsymbol{\omega}_0 = \mathbf{0}$ of Eq. (19), $a(0) = \dot{\boldsymbol{\omega}}_0 = \mathbf{0}$ of Eq. (21)), and the parallel evaluation of Eq. (23) or Eq. (31) or Eq. (33) requires $3\lceil \log_2(n+1) \rceil$ scalar additions for the case of $a(0) \neq \mathbf{0}$ (i.e., $a(0) = \dot{\mathbf{v}}_0$ of Eq. (23), $a(0) = \mathbf{f}_e$ of Eq. (31), $a(0) = \mathbf{n}_e$ of Eq. (33)), then the total evaluation of these equations requires $(6\lceil \log_2 n \rceil + 9\lceil \log_2(n+1) \rceil)$ scalar additions.

(c) The parallel evaluation of the other equations in the NE formulation, e.g., $\mathbf{p}_i^* = {}^0\mathbf{R}_i \, {}^i\mathbf{p}_i^*$, $\bar{\mathbf{s}}_i = {}^0\mathbf{R}_i \, {}^i\bar{\mathbf{s}}_i$, $\bar{\mathbf{a}}_i$, $\mathbf{F}_i$, $\mathbf{N}_i$, $\boldsymbol{\tau}_i$, and all the b_i of Eqs. (18), (20), (22), and (32) can be calculated by simple parallel computations, yielding a constant computation of 135 scalar multiplications and 98 scalar additions.

By combining the results of (a), (b), and (c), the total computational complexity of the parallel algorithm applied to an n-link PUMA-like robot can be given as follows: $\{27\lceil \log_2 n \rceil + 135\}$ scalar multiplications and $\{24\lceil \log_2 n \rceil + 9\lceil \log_2(n+1) \rceil + 98\}$ scalar additions. If $n = 6$, then the complexity of the parallel NE algorithm is 216 mults and 197 adds as compared with the complexity of the NE algorithm running on a uniprocessor (see Appendix B): 678 mults and 597 adds. Thus, the speedup for multiplication is $678/216 = 3.14$ and for addition is $597/197 = 3.03$. Moreover, even if n becomes large, say $n = 12$ (for redundant robots), then the number of multiplications and additions increases only by 27 and 33, respectively. The complexity analysis of the above parallel Newton-Euler is tabulated in Table 2.

The above complexity analysis shows that the bottleneck of parallel computation of the inverse dynamics depends on solving the homogeneous linear recurrence of the NE formulation. Furthermore, the above parallel algorithm requires that one processor "handles" one joint. It is desirable to relax this strict requirement and obtain an efficient parallel algorithm which can greatly improve the evaluation of the linear recurrence using p processors. A parallel algorithm of evaluating the inverse dynamics with a restricted number of p processors, $1 \le p \le n$, to achieve the time lower bound of $O(k_1\lceil n/p \rceil + k_2\lceil \log_2 p \rceil)$ will be discussed in the next subsection.

3.4. Efficient p-Fold Parallel Algorithm

This subsection describes an efficient p-fold parallel algorithm for evaluating the inverse dynamics using a restricted number of p processors, $1 < p \le n$, to achieve the time lower bound of $O(k_1\lceil n/p \rceil + k_2\lceil \log_2 p \rceil)$. The p-fold parallel algorithm can be best described as consisting of p-parallel blocks with pipelined elements within each parallel block. The results from the computations in the p blocks form a new homogeneous linear recurrence of size p, which again can be computed using the recursive doubling algorithm. The parallel algorithm with p-fold parallelism (PFP) is summarized and presented as follows.

Table 2. Breakdown of Mathematical Operations of the Parallel Newton-Euler Computation for an n-link PUMA-like Manipulator.

Steps	Variables	Multiplications	Additions
Step 1	${}^0\mathbf{R}_i$	$27\lceil \log_2 n \rceil$	$18\lceil \log_2 n \rceil$
Step 2	$\mathbf{p}_i^*$	9	6
	$\bar{\mathbf{s}}_i$	9	6
Step 3	b_i	3	0
	ω_i	0	$3\lceil \log_2 n \rceil$
Step 4	b_i	9	6
	$\dot{\omega}_i$	0	$3\lceil \log_2 n \rceil$
Step 5	b_i	18	12
	$\dot{\mathbf{v}}_i$	0	$3\lceil \log_2 (n+1) \rceil$
Step 6	$\bar{\mathbf{a}}_i$	18	15
Step 7	$\mathbf{F}_i$	3	0
Step 8	${}^i\omega_i$	9	6
	${}^i\dot{\omega}_i$	9	6
	${}^i\mathbf{N}_i$	24	18
	$\mathbf{N}_i$	9	6
Step 9	$\mathbf{f}_i$	0	$3\lceil \log_2 (n+1) \rceil$
Step 10	b_i	12	15
	$\mathbf{n}_i$	0	$3\lceil \log_2 (n+1) \rceil$
Step 11	τ_i	3	2
Total Number of Mathematical Operations		$27\lceil \log_2 n \rceil + 135$ (216, when $n = 6$)	$24\lceil \log_2 n \rceil + 9\lceil \log_2 (n+1) \rceil + 98$ (197, when $n = 6$)

Algorithm PFP (p–fold Parallelism). This algorithm divides the computations into p-parallel blocks of computations. The jth processor computes the elements in the jth block serially. The results from the the p-parallel blocks form a new homogeneous linear recurrence of size p, which can be computed by the recursive doubling algorithm.

P1. [Initialization.] Given $a(0) \neq$ identity, let $s = \lceil (n+1)/p \rceil$, $m = ps$, and set

$$M(i) = \begin{cases} a(i-1), & 1 \le i \le n+1 \\ 0, & n+1 < i \le m \end{cases} \tag{35}$$

where s indicates the block size (i.e., number of elements in a block).

P2. [Divide into p blocks.] Divide $M(i)$, $1 \le i \le m$, into p blocks as follows:
jth block $= \{M((j-1)s+1),\ M((j-1)s+2),\ \cdots, M(js)\}$,
and let $N(1,j) = M((j-1)s+1)$, $1 \le j \le p$,
where $N(i,j)$ indicates the ith element in the jth block.

P3. [Compute $N(i,j)$ in a do-loop.] The jth processor serially computes $N(i,j)$ in the jth block as

for $i \leftarrow 2$ **to** s **do**

$$N(i,j) = N(i-1,j) * M((j-1)s+i) \ , \quad 1 \le j \le p \tag{36}$$

end_do

$x(1), x(2), \cdots, x(s-1)$ are evaluated in the do-loop as well as $N(i,1)$, $2 \le i \le s$. That is, $x(1) = N(2,1)$, $x(2) = N(3,1)$, $\cdots$, $x(s-1) = N(s,1)$.

P4. [Form a new homogeneous linear recurrence of size p.] Let $\gamma(j) = N(s,j)$ and $y(j) = x(js-1)$ for $1 \le j \le p$. Referring to Eqs. (10), (35), and (36), we have

$$\begin{aligned} y(j) &= x(js-1) \\ &= x((j-1)s-1) * a((j-1)s) * a((j-1)s+1) * \cdots * a(js-1) \\ &= y(j-1) * M((j-1)s+1) * M((j-1)s+2) * \cdots * M(js) \\ &= y(j-1) * N(1,j) * M((j-1)s+2) * \cdots * M(js) \\ &= y(j-1) * N(s,j) \\ &= y(j-1) * \gamma(j) . \end{aligned} \tag{37}$$

Equation (37) is a new homogeneous linear recurrence of size p which can be parallelly evaluated by the algorithm FOHRA, running in time proportional to $O(\lceil \log_2 p \rceil)$ and yielding the results $x(s-1) = y(1)$, $x(2s-1) = y(2)$, $\cdots$, $x(ps-1) = y(p)$. Note that if $(n+1)$ is divisible by p, then $x(n) = y(p)$; otherwise,

$x(ps-1) = y(p) = 0.$

P5. [Compute intermediate $x(i)$.] Without loss of generality, assuming that $(n+1)$ is not divisible by p, then there are $n-p-s+3$ intermediate terms of $x(i)$ that need to be determined. They are $x(js+i-1)$, $1 \le i \le s-1$, $1 \le j \le p-2$, and $x((p-1)s+i-1)$, $1 \le i \le (n+1)-(p-1)s$.

Referring to Eqs. (11), (36), and (37), thereby giving

$$\begin{aligned} x(js+i-1) &= x(js-1) * a(js) * a(js+1) * \cdots * a(js+i-1) \\ &= y(j) * M(js+1) * M(js+2) * \cdots * M(js+i) \\ &= y(j) * N(i,j+1) \quad , \quad 1 \le i \le s-1,\ 1 \le j \le p-2 \end{aligned} \tag{38}$$

$$\text{and} \quad x((p-1)s+i-1) = y(p-1) * N(i,p) \quad , \quad 1 \le i \le (n+1)-(p-1)s$$

where $N(i,j)$, $y(j)$ of Eq. (38) have been evaluated in steps P3 and P4, respectively. Equation (38) shows that if $(n-p-s+3)$ tasks (terms) are of the same evaluation, then the calculations of these equal tasks are suited to an SIMD computer of p processors. It is shown that parallel evaluation of Eq. (38) requires $\lceil (n-p-s+3)/(p-1) \rceil$ time steps. Note that if $(n+1)$ is divisible by p, then $x(n)$ can be evaluated in step P4, yielding $(n-p-s+2)$ equal tasks in Eq. (38), thereby requiring $\lceil (n-p-s+2)/(p-1) \rceil$ time steps.

END PFP

It is seen that the total parallel computing time T_p of the homogeneous linear recurrence of size $(n+1)$ using p processors is:

$$T_p = \begin{cases} \lceil (n+1)/p \rceil + \lceil (n-p-s+3)/(p-1) \rceil + \lceil \log_2 p \rceil - 1, & \text{if } (n+1) \text{ is not divisible by } p. \\ \lceil (n+1)/p \rceil + \lceil (n-p-s+2)/(p-1) \rceil + \lceil \log_2 p \rceil - 1, & \text{if } (n+1) \text{ is divisible by } p. \end{cases} \tag{39}$$

Applying the above p-fold parallel algorithm to the NE formulation for an n-link PUMA-like manipulator, it is able to achieve the time lower bound of $O(k_1 \lceil n/p \rceil + k_2 \lceil \log_2 p \rceil)$. As an example, using Table 2 and by assuming n and $(n+1)$ are not divisible by p, (note that Eqs. (14), (19), and (21), are linear recurrence of size n while Eqs. (23), (31), and (33) are linear recurrence of size $(n+1)$), the p-fold parallelism requires $\{ 27\lceil \log_2 p \rceil + 162\lceil n/p \rceil + 27\lceil (n-p-s_1+1)/(p-1) \rceil - 27 \}$ scalar mults and $\{ 33\lceil \log_2 p \rceil + 122\lceil n/p \rceil + 9\lceil (n+1)/p \rceil + 24\lceil (n-p-s_1+1)/(p-1) \rceil + 9\lceil (n-p-s_2+2)/p \rceil - 33 \}$ scalar adds, where $s_1 = \lceil n/p \rceil$ and $s_2 = \lceil (n+1)/p \rceil$, to evaluate per complete set of joint torques.

Though the above p-fold parallel algorithm achieves the time lower bound on computing the inverse dynamics, it is desirable to determine the minimum number of processors that can execute the p-fold parallel algorithm and achieve the time lower bound. In order to investigate the trade-off between computing time and the number of processors, the following lemma will be utilized:

Lemma 2: The lower bound of time-processor product $p \cdot T_p\,[E<n>]$ [40]. Given a linear arithmetic expression $E<n>$, its associated parallel computing time $T_p\,[E<n>]$, and p processors, we have

$$p \cdot T_p\,[E<n>] \geq O(n).$$

The above lemma can be easily proved using Lemma 1. From Eqs. (6) and (7), we see that since $T_p\,[\tau_1, \tau_2, \cdots, \tau_n] \geq T_p\,[E<n>]$, we have $p \cdot T_p\,[\tau_1, \tau_2, \cdots, \tau_n] \geq O(n)$. If $p = O(n/\lceil \log_2 n \rceil)$ processors are used, then $p \cdot T_p\,[\tau_1, \tau_2, \cdots, \tau_n] = O(n)$ can be achieved. Hence, with $p = O(n/\lceil \log_2 n \rceil)$ processors, the parallel computing time $T_p\,[\tau_1, \tau_2, \cdots, \tau_n]$ achieves the time lower bound of $O(\lceil \log_2 n \rceil)$. This suggests that if one uses $p = O(n/\lceil \log_2 n \rceil)$ processors in the p-fold parallelism, then the parallel computing time $T_p\,[\tau_1, \tau_2, \cdots, \tau_n]$ of the p-fold parallel algorithm achieves the time lower bound of $O(\lceil \log_2 n \rceil)$. This also implies that a logarithmic factor reduction in the number of processors, from n to $n/\lceil \log_2 n \rceil$, will only result in a constant factor increase in parallel computing time, and the resulting implementation is optimal with respect to their usage of parallel computing time and the number of processors.

3.5. Interconnections Between Processors

The above n-fold parallel algorithm is well suited to be run on an SIMD computer. The physical structure of an SIMD computer can be viewed as a set of n processing elements (PEs), where each PE consists of a processor with its own memory and the operations performed by each processor involve at most two operands. A cascade structure was suggested for connecting the PEs (Algorithm FOHRA). An alternate structure is to position a network between the processors and memories. If a complete interconnection network is used, every processor is directedly connected to every other processor and requires $n(n-1)/2$ bi-directional links between n processors; while the cascade network shown in Figure 6 requires $(n+1)(\lceil \log_2 (n+1) \rceil - 1) + 1$ links among n processors (note that the null operation in Figure 6 indicates no connection). Both cascade and complete networks have difficulty on their complex structure of linkage and interconnection. It is more desirable to have a simple network structure with fewer interconnection paths and the number of links between processors is proportional to n [41-45]. The interconnection pattern, called the "perfect shuffle" [46,47,41] and shown in Figure 7, has such properties. The pattern shifts data cyclically among processors and routes data by a permutation function. A careful study of Figure 7 shows that the indices on the left are mapped onto the indices on the right according to the permutation function $P(i)$

$$P(i) = \begin{cases} 2i, & 0 \leq i \leq (N/2) - 1 \\ 2i + 1 - N, & N/2 \leq i \leq N - 1 \end{cases} \tag{40}$$

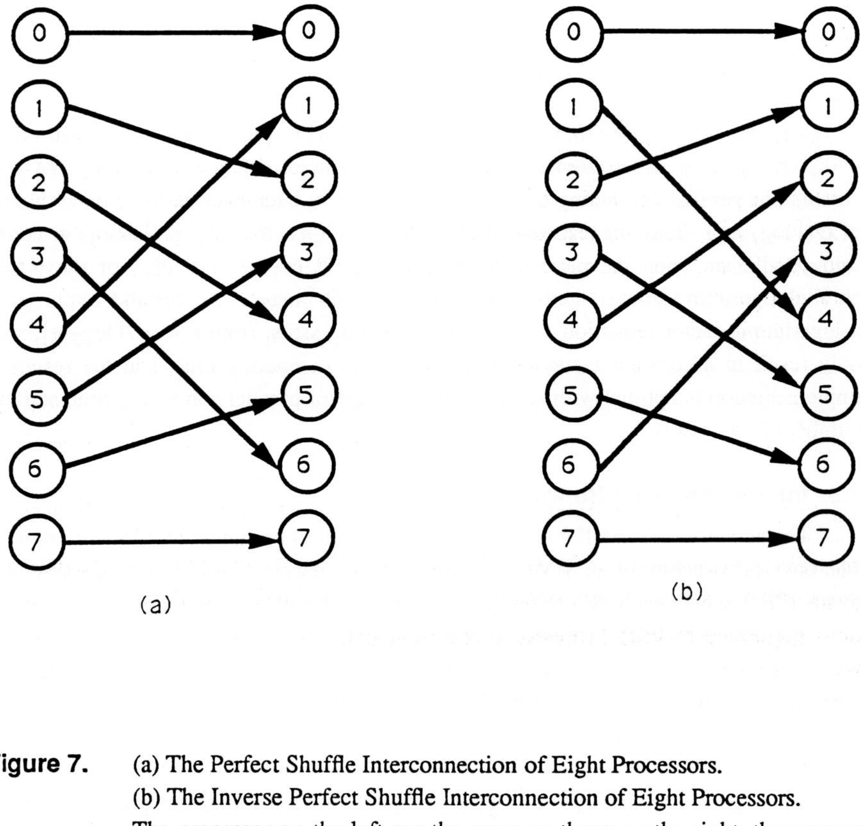

Figure 7. (a) The Perfect Shuffle Interconnection of Eight Processors.
(b) The Inverse Perfect Shuffle Interconnection of Eight Processors.
The processor on the left are the same as those on the right, they appear twice for clarity.

where N is an even number and equal to $(n+1)$. If N is an odd number, a null element can be placed to the excess index, making it an even number.

An attractive interconnection pattern, called "inverse perfect shuffle" [46,47], is suitable for the implementation of solving the homogeneous linear recurrence and can be obtained by reversing the arrows of the perfect shuffle in Figure 7. To show how the inverse perfect shuffle is used, an example of evaluating a recursive vector addition problem of size $(n+1)$ is stated as follows: Given a sequence of vectors $a(i)$, $(0 \le i \le n,\ n = (N-1))$, find $y(1), y(2), \cdots, y(n)$, where

$$y(i) = y(i-1) + a(i). \tag{41}$$

The parallel computation scheme of Eq. (41) is shown in Figure 6 (where the operator "*" is replaced by "+"). The computation scheme can be implemented in the computer organization as shown in Figure 8, using the inverse perfect shuffle. The open circled nodes of Figure 6 do not perform any computation (i.e., null operation), but merely pass through a single operand. A mask is introduced to indicate a null operation as

$$\text{mask} = \begin{cases} 1, & (Y_{i,j}\text{ , if } i \ne 0) \\ 0, & (Y_{i,j}\text{ , if } i = 0) \end{cases}$$

where $Y_{i,j}$ indicates the summation $\sum_{k=i}^{j} a(k)$. It is not difficult to show that $Y_{0,j} (= \sum_{k=0}^{j} a(k))$ locates on the open circled node of Figure 6 and requires a null operation by setting the mask to "0". A general flow of describing the operation through the computer organization can be described by the following INVSHU algorithm.

Algorithm INVSHU (Inverse Shuffle Scheme).

I1. [Initialization.] Set $Y(i) \leftarrow a(i)$ for $0 \le i \le n$ and $n \leftarrow N-1$.

I2. [Compute masks.]

```
for i ← 1 to ⌈log (n+1)⌉ do
    Y(i) ← Y(i) + Y(i−1),  (mask)
    Inverse shuffle (Y)
    Compute new mask
end_do
```

I3. [Output $y(i)$.] Decide the index i^* of the resulting $Y(i^*)$ corresponds to $y(i)$, and set $y(i) \leftarrow Y(i^*)$. If N is equal to 2^L, then $i^* = i$, where L is any integer.

END INVSHU

In the first iteration, the sums computed are sums of adjacent elements of $a(i)$. That is, $Y(i) = a(i) + a(i-1)$. After the inverse shuffle, the items that are adjacent in $Y(i)$ are

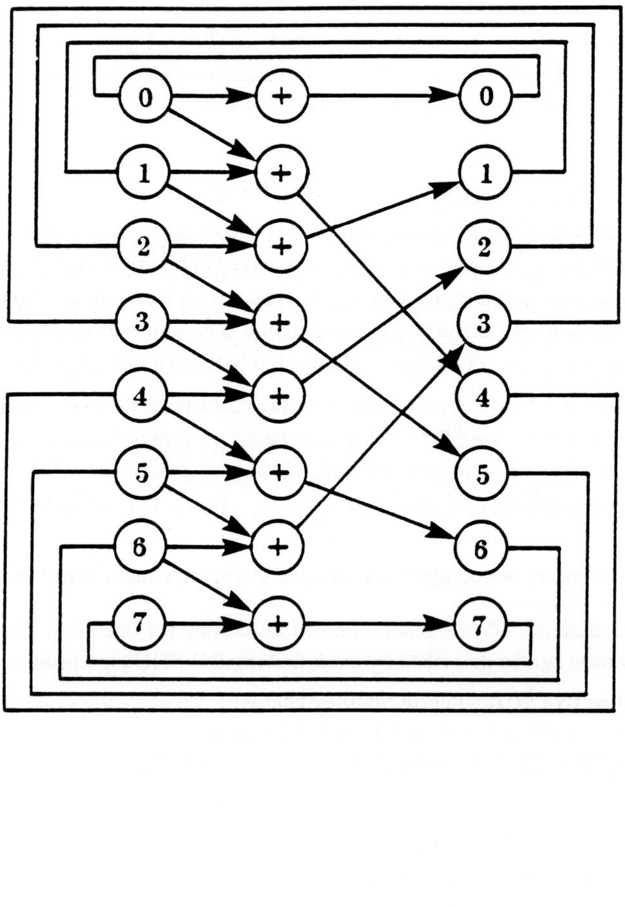

Figure 8. Data Flow for Recursive Doubling Algorithm. The "+" can be any associate operator.

pairs of sums of elements of $a(i)$, and, in fact, are $a(i)+a(i-1)$ and $a(i-2)+a(i-3)$ in adjacent words, except in a few elements that represent special cases. Table 3 shows the corresponding values of i^* for various values of N. Table 4 shows the intermediate results in the computation, and the intermediate values of the mask bits. It is rather interesting to see that the mask bits for one iteration can be computed from the mask bits from the preceding calculation by first replicating the present mask bits with a cyclic shift, followed by passing the resulting vector through the inverse perfect shuffle network.

3.6. VLSI Implementation

The parallel NE equations formulated as a homogeneous recurrence problem (Eqs. (14)-(34)) are also susceptible to a VLSI chip implementation. The essential primitive processing elements of the VLSI chip can be constructed by a group of Modular Processors (MPs), which have a microprocessor-like architecture [27]. It is being used to evaluate the operations of two 3×1 vectors, i.e., vector dot product or vector addition, consisting of three scalar processing units that can perform scalar addition or multiplication on each coordinate of those two 3×1 vectors concurrently. It is not difficult to show that the parallel processing time of evaluating 3×1 vector dot product and 3×1 vector addition takes 1 mult and 2 adds (or 3 Flops) and 1 add (or 1 Flop), respectively. Here we assume 1 mult = 1 add = 1 floating-point operation (Flop)

Three processing elements are most useful for the VLSI implementation of the parallel NE equations:

(1) A Vector Cross Product (VCP) processing element is composed of 3 MPs that perform concurrently and can be used to parallelly evaluate the cross product of two 3×1 vectors, taking 1 mult and 1 add (or 2 Flops).

(2) A Matrix Vector Product (MVP) processing element is composed of 3 MPs that perform concurrently and can be used to evaluate the product between a 3×3 matrix and a 3×1 vector, taking 1 mult and 2 adds (or 3 Flops).

(3) A Matrix Matrix Product (MMP) processing element is composed of 3 MVPs (or 9 MPs) that perform concurrently and can be used to parallelly evaluate the product of two 3×3 matrices, taking 1 mult and 2 adds (or 3 Flops).

Though there are many different approaches to the design of VLSI implementation [48-51], the systolic pipelined structure may be the most suitable for implementing the parallel NE algorithm. The systolic pipelined machine represents one class of such machines that have regular structure of primitive processing cells. Each cell regularly pumps data in and out by simple control and communication [51]. It is important to show that the processing time of each cell (or subcell) must make compromise with the same time steps as close as possible, so that a regular flow of data is kept up in the

Table 3. The corresponding i^* for $N = 6, 10, 12, 14, 18, 20, 22, 24, 26, 28, 30$.

i^* \ i	0	1	2	3	4	5	6	7	8	9	10	11	12	13	14	15	16	17	18	19	20	21	22	23	24	25	26	27	28	29
N=6	0	3	1	4	2	5																								
N=10	0	7	5	3	1	8	6	4	2	9																				
N=12	0	5	10	4	9	3	8	2	7	1	6	11																		
N=14	0	3	6	9	12	2	5	8	11	1	4	7	10	13																
N=18	0	15	13	11	9	7	5	3	1	16	14	12	10	8	6	4	2	17												
N=20	0	13	7	1	14	8	2	15	9	3	16	10	4	17	11	5	8	12	6	19										
N=22	0	11	1	12	2	13	3	14	4	15	5	16	6	17	7	18	8	19	9	20	10	21								
N=24	0	9	18	4	13	22	8	17	3	12	19	7	16	2	11	20	6	15	1	10	19	5	14	23						
N=26	0	7	14	21	3	10	17	24	6	13	20	2	9	16	23	5	12	19	1	8	15	22	4	11	18	25				
N=28	0	5	10	15	20	25	3	8	13	18	23	1	6	11	16	21	26	4	9	14	19	24	2	7	12	17	22	27		
N=30	0	3	6	9	12	15	18	21	24	27	1	4	7	10	13	16	19	22	25	28	2	5	8	11	14	17	20	23	26	29

Table 4. Intermediate values for recursive doubling calculation of $y(i) = y(i-1) + a(i)$. Note that $Y_{i,j}$ denotes $\sum_{k=i}^{j} a(k)$.

Y	Mask	Sum	Inverse Shuffle	Mask	Sum	Inverse Shuffle	Mask	Sum	Inverse Shuffle
$a(0)=Y_{0,0}$	0	$Y_{0,0}$	$Y_{0,0}$	0	$Y_{0,0}$	$Y_{0,0}$	0	$Y_{0,0}$	$Y_{0,0}=y(0)$
$a(1)=Y_{1,1}$	1	$Y_{0,1}$	$Y_{1,2}$	1	$Y_{0,2}$	$Y_{1,4}$	1	$Y_{0,4}$	$Y_{0,1}=y(1)$
$a(2)=Y_{2,2}$	1	$Y_{1,2}$	$Y_{3,4}$	1	$Y_{1,4}$	$Y_{0,1}$	0	$Y_{0,1}$	$Y_{0,2}=y(2)$
$a(3)=Y_{3,3}$	1	$Y_{2,3}$	$Y_{5,6}$	1	$Y_{3,6}$	$Y_{2,5}$	1	$Y_{0,5}$	$Y_{0,3}=y(3)$
$a(4)=Y_{4,4}$	1	$Y_{3,4}$	$Y_{0,1}$	0	$Y_{0,1}$	$Y_{0,2}$	0	$Y_{0,2}$	$Y_{0,4}=y(4)$
$a(5)=Y_{5,5}$	1	$Y_{4,5}$	$Y_{2,3}$	1	$Y_{0,3}$	$Y_{3,6}$	1	$Y_{0,6}$	$Y_{0,5}=y(5)$
$a(6)=Y_{6,6}$	1	$Y_{5,6}$	$Y_{4,5}$	1	$Y_{2,5}$	$Y_{0,3}$	0	$Y_{0,3}$	$Y_{0,6}=y(6)$
$a(7)=Y_{7,7}$	1	$Y_{6,7}$	$Y_{6,7}$	1	$Y_{4,7}$	$Y_{4,7}$	1	$Y_{0,7}$	$Y_{0,7}=y(7)$

systolic pipelined array. Otherwise, delay buffers must be inserted along various paths to achieve maximum pipelining [52]. The systolic pipelined implementation of the parallel NE algorithm for a manipulator with all rotary joints is shown in Figure 9. The primitive processing element assignment and evaluation arrangement of each cell of Figure 9 have been suggested below, which are used to keep the data flow of the systolic pipelined array fluently and improve the utilization of the VLSI implementation. Referring to Figure 9 and Eqs. (14)-(34) of the NE equations of motion in linear recurrence form, we can give the relationship between the cells of Figure 9 and the corresponding equations of the NE algorithm as follows:

(a) Cells 1, 4, 6, 9, 15, and 17 are used to evaluate the homogeneous linear recurrences of ${}^0\mathbf{R}_i$ in Eq. (14), $\boldsymbol{\omega}_i$ in Eq. (19), $\dot{\boldsymbol{\omega}}_i$ in Eq. (21), $\dot{\mathbf{v}}_i$ in Eq. (23), $\mathbf{f}_i$ in Eq. (31), and $\mathbf{n}_i$ in Eq. (33), respectively. Cell 1 is implemented as a cascade network of $n(\lceil \log_2 n \rceil - 1) + 1$ MMPs shown in Figure 6, and taking 1 mult and 2 adds (or 3 Flops) on each node (or subcell) of the network. Cell 4 or cell 6 (or cell 9, cell 15, cell 17) is implemented as a cascade network of $n(\lceil \log_2 n \rceil - 1)$ (or (n+1) $(\lceil \log_2 (n+1) \rceil - 1) + 1$) MPs, and taking 1 add (or 1 Flop) on each node (or subcell) of the network.

(b) Cells 3, 5, 7 and 8, and 16 are used to evaluate the b_i of the homogeneous linear recurrences in Eqs. (18), (20), (22), and (32), respectively. The processor assignment and evaluation arrangement can be shown in detail as follows:

(i) Cell 3 is composed of n MPs and takes 1 mult to evaluate Eq. (18).

(ii) Cell 5 is composed of n MPs and n VCPs. n MPs evaluate $\mathbf{z}_{i-1}\ddot{q}_i$ and n VCPs evaluate $\boldsymbol{\omega}_{i-1} \times \mathbf{z}_{i-1}\dot{q}_i$ (note that $\mathbf{z}_{i-1}\dot{q}_i$ has been evaluated in cell 3) concurrently, then $b_i = \mathbf{z}_{i-1}\ddot{q}_i + \boldsymbol{\omega}_{i-1} \times \mathbf{z}_{i-1}\dot{q}_i$ is evaluated. The total processing time of cell 5 amounts to 1 mult and 2 adds (or 3 Flops).

(iii) Cell 7 is used to evaluate $\mathbf{t3}_i = \boldsymbol{\omega}_i \times (\boldsymbol{\omega}_i \times \mathbf{p}_i^*)$. It is divided into two subtasks, each of them consists of n VCPs. The first subcell is used to evaluate $\mathbf{t1}_i = \boldsymbol{\omega}_i \times \mathbf{p}_i^*$, and the second subcell is used to evaluate $\mathbf{t2}_i = \boldsymbol{\omega}_i \times \mathbf{t1}_i$, both subcells take 1 mult and 1 add (or 2 Flops), respectively. Cell 8 is composed of n VCPs and used to evaluate $b_i = \mathbf{t2}_i + \dot{\boldsymbol{\omega}}_i \times \mathbf{p}_i^*$ of Eq. (22), taking 1 mult and 2 adds (or 3 Flops).

(iv) Cell 16 is divided into three subcells. The first subcell is composed of n VCPs and used to evaluate $\mathbf{t4}_i = (\mathbf{p}_i^* + \bar{\mathbf{s}}_i) \times \mathbf{F}_i$, taking 1 mult and 2 adds (or 3 Flops). The second subcell is also composed of n VCPs and used to evaluate $\mathbf{t5}_i = \mathbf{t4}_i + \mathbf{p}_i^* \times \mathbf{f}_{i+1}$, taking 1 mult and 2 adds (or 3 Flops). The third subcell is composed of n MPs and used to evaluate $b_i = \mathbf{t5}_i + \mathbf{N}_i$, taking 1 add (or 1 Flop).

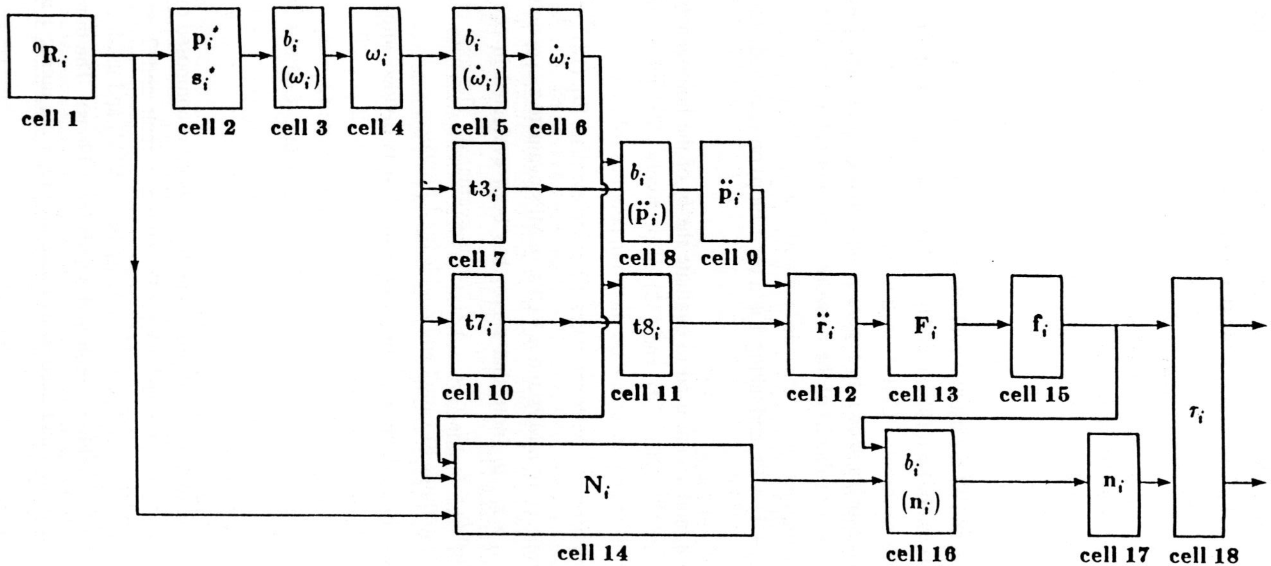

Figure 9. VLSI Implementation of Parallel Newton-Euler Algorithm for a Manipulator with Rotary Joints.

(c) Cell 2 is composed of $2n$ MVPs and used to evaluate $\mathbf{p}_i^*$ of Eq. (16) and $\overline{\mathbf{s}}_i$ of Eq. (15) concurrently, taking 1 mult and 2 adds (or 3 Flops).

(d) Cells 10, 11, and 12 are used to evaluate $\overline{\mathbf{a}}_i$ of Eq. (24).

(i) Cell 10 is divided into two subcells. The first subcell is composed of n VCPs and used to evaluate $\mathbf{t6}_i = \boldsymbol{\omega}_i \times \overline{\mathbf{s}}_i$, taking 1 mult and 1 add (or 2 Flops). The second subcell is also composed of n VCPs and used to evaluate $\mathbf{t7}_i = \boldsymbol{\omega}_i \times \mathbf{t6}_i$, taking 1 mult and 1 add (or 2 Flops).

(ii) Cell 11 is composed of n VCPs and used to evaluate $\mathbf{t8}_i = \dot{\boldsymbol{\omega}}_i \times \overline{\mathbf{s}}_i + \mathbf{t7}_i$, taking 1 mult and 2 adds (or 3 Flops).

(iii) Cell 12 is composed of n MPs and used to evaluate $\overline{\mathbf{a}}_i = \mathbf{t8}_i + \dot{\mathbf{v}}_i$, taking 1 add (or 1 Flop).

(e) Cell 13 is composed of n MPs and used to evaluate $\mathbf{F}_i = m_i\, \overline{\mathbf{a}}_i$, taking 1 mult (or 1 Flop).

(f) Cell 14 is used to evaluate $\mathbf{N}_i$ and is divided into 6 subcells as follows:

(i) The first subcell is composed of n MVPs and used to evaluate ${}^i\boldsymbol{\omega}_i$ of Eq. (27), taking 1 mult and 2 adds (or 3 Flops).

(ii) The second subcell is composed of n MVPs and used to evaluate ${}^i\dot{\boldsymbol{\omega}}_i$ of Eq. (28), taking 1 mult and 2 adds (or 3 Flops).

(iii) The third, fourth, and fifth subcells are used to evaluate ${}^i\mathbf{N}_i$ of Eq. (29). The third subcell is composed of n MVPs and used to evaluate $\mathbf{t9}_i = {}^i\mathbf{I}_i\, {}^i\boldsymbol{\omega}_i$, taking 1 mult and 2 adds (or 3 Flops). The fourth subcell is composed of n MVPs and used to evaluate $\mathbf{t10}_i = {}^i\mathbf{I}_i\, {}^i\dot{\boldsymbol{\omega}}_i$, taking 1 mult and 2 adds (or 3 Flops). The fifth subcell is composed of n VCPs and used to evaluate ${}^i\mathbf{N}_i = \mathbf{t10}_i + {}^i\boldsymbol{\omega}_i \times (\mathbf{t9}_i)$, taking 1 mult and 2 adds (or 3 Flops).

(iv) The sixth subcell is composed of n MVPs and used to evaluate $\mathbf{N}_i$ of Eq. (30) taking 1 mult and 2 adds (or 3 Flops).

(g) Cell 18 is composed of n MPs and used to evaluate $\tau_i(t)$ of Eq. (34), taking 1 mult and 2 adds (or 3 Flops).

In order to evaluate the performance of the above VLSI architecture, two performance measures, the initial delay time performance (or settling time) and the pipelined time performance, are utilized. The initial delay time performance is defined as the required time interval for obtaining the first resulting generalized force/torque vector by the systolic pipelined implementation. The pipelined time performance is used to measure the throughput of the systolic pipeline implementation, and being defined as the time steps in the interval between successive input generalized force/torque vector $\boldsymbol{\tau}(t)$ by the pipeline system. Observing the processing time of each subcell (or cell) of the pipeline system implementation as shown in Figure 9, it is seen that the initial delay time depends on the processing time for executing the tasks, which pass through the subpaths as shown in Figure 9. By counting the processing time of each subpath, it is not difficult to show that the subpath passing through Cells 1-6, Cell 14, and Cells 16-18 with a dominant processing time of $(\lceil \log_2 n \rceil + 12)$ mults and $(4\lceil \log_2 n \rceil + \lceil \log_2(n+1) \rceil + 23)$ adds (or $(5\lceil \log_2 n \rceil + \lceil \log_2(n+1) \rceil + 35)$ Flops) for $n \leq 8$, and the subpath passing through Cells 1-9, Cells 12-13, Cells 15-18 with a dominant processing time of $(\lceil \log_2 n \rceil + 11)$ mults and $(4\lceil \log_2 n \rceil + 3\lceil \log_2(n+1) \rceil + 18)$ adds (or $(5\lceil \log_2 n \rceil + 3\lceil \log_2(n+1) \rceil + 29)$ Flops) for $n > 8$. Moreover, the pipelined time is the maximum processing time of each subcell (or cell); that is, 1 mult and 2 adds (or 3 Flops), requiring $\{ n(\lceil \log_2 n \rceil - 1) + 1 \}$ MMPs, $7n$ MVPs, $10n$ VCPs, and $\{ 2n(\lceil \log_2 n \rceil) + 3(n+1)(\lceil \log_2(n+1) \rceil) + n+2 \}$ MPs (or $\{ 11n(\lceil \log_2 n \rceil) + 3(n+1)(\lceil \log_2(n+1) \rceil) + 43n + 11 \}$ MPs). If $n = 6$, then the VLSI implementation requires 530 MPs with an initial delay time of 15 mults and 38 adds (or 53 Flops) and a pipelined time of 1 mult and 2 adds (or 3 Flops).

The VLSI implementation shows that as the number of processors used increases, the computing time of evaluating successive applied generalized force vector will be drastically reduced and independent of the number of links. The results also show that the parallel NE formulation is susceptible to a systolic pipelined architecture implementation with an initial delay time of 53 Flops and a pipelined time of 3 Flops for a 6-link manipulator with all rotary joints. This yields the ability in computing the inverse dynamics of a manipulator in real time.

4. CONCLUSION

In this chapter, a cost-effective, microprocessor-based, pipelined architecture, parallel algorithms, and a VLSI systolic pipeline were developed to compute the inverse dynamics for the real-time control of a manipulator. The analysis of the dynamic equations of motion of a manipulator lead us to a pipelined architecture based on the computational flow of the NE equations of motion. The proposed pipelined architecture consists of currently available microprocessors arranged in such a way as to permit pipelining at a macro level, with parallelism within each macro block. To further speed up the computations and reduce the order of computational complexity, a novel parallel algorithm for computing the inverse dynamics using the NE equations of motion was developed to be implemented on an SIMD computer with p processors to achieve the time lower bound of $O(k_1 \lceil n/p \rceil + k_2 \lceil \log_2 p \rceil)$, where k_1 and k_2 are constants. When $p = n$, the proposed p-fold parallel algorithm achieves the time lower bound of $O(\lceil \log_2 n \rceil)$, and can be best described as consisting of p-parallel blocks with pipelined elements within each parallel block. The interconnection networks for the processors have also been investigated to improve the utilization of communication and internal buffering between processors in an SIMD computer. The inverse perfect shuffle interconnection scheme was found to be superior and its number of links between the processors is proportional to the number of processors. Furthermore, the proposed parallel algorithm is susceptible to a systolic pipelined architecture, with an initial delay time of 15 mults and 38 adds (or 53 Flops) and a pipelined time of 1 mult and 2 adds (or 3 Flops), yielding the ability to rapidly evaluate the successive sets of joint torques of a manipulator in real time.

Appendix A
Euler-Lagrange Equations of Motion for a Manipulator with Rotary Joints

$$\tau_i = \sum_{k=i}^{n} \sum_{j=1}^{k} Tr\left\{ \frac{\partial\, {}^0\mathbf{T}_k}{\partial\theta_j} \mathbf{J}_k \left[\frac{\partial\, {}^0\mathbf{T}_k}{\partial\theta_i} \right]^T \right\} \ddot{\theta}_j + \sum_{r=i}^{n} \sum_{j=1}^{r} \sum_{k=1}^{r} Tr\left\{ \frac{\partial^2\, {}^0\mathbf{T}_r}{\partial\theta_j \partial\theta_k} \mathbf{J}_r \left[\frac{\partial\, {}^0\mathbf{T}_r}{\partial\theta_i} \right]^T \right\} \dot{\theta}_j \dot{\theta}_k$$

$$+ \sum_{j=i}^{n} m_j \mathbf{g} \left\{ \frac{\partial\, {}^0\mathbf{T}_j}{\partial\theta_i} \right\} \bar{\mathbf{r}}_j \quad , \quad i = 1, 2, \cdots, n \tag{A-1}$$

or expressed in matrix form explicitly:

$$\tau_i = \sum_{k=1}^{n} D_{ik} \ddot{\theta}_k + \sum_{k=1}^{n} \sum_{m=1}^{n} H_{ikm} \dot{\theta}_k \dot{\theta}_m + G_i \quad , \quad i = 1, 2, \cdots, n \tag{A-2}$$

where

$$D_{ik} = \sum_{j=max(i,k)}^{n} Tr\left\{ \frac{\partial\, {}^0\mathbf{T}_j}{\partial\theta_k} \mathbf{J}_j \left[\frac{\partial\, {}^0\mathbf{T}_j}{\partial\theta_i} \right]^T \right\} \tag{A-3}$$

$$H_{ikm} = \sum_{j=max(i,k,m)}^{n} Tr\left\{ \frac{\partial^2\, {}^0\mathbf{T}_j}{\partial\theta_k \partial\theta_m} \mathbf{J}_j \left[\frac{\partial\, {}^0\mathbf{T}_j}{\partial\theta_i} \right]^T \right\} \tag{A-4}$$

$$G_i = \sum_{j=i}^{n} \left[m_j \mathbf{g} \left\{ \frac{\partial\, {}^0\mathbf{T}_j}{\partial\theta_i} \right\} \bar{\mathbf{r}}_j \right] \tag{A-5}$$

and where

τ_i generalized torque for joint i

θ_i joint displacement of joint i

$\dot{\theta}_i$ joint velocity of joint i

$\ddot{\theta}_i$ joint acceleration of joint i

Tr Trace operator

${}^i\mathbf{T}_j$ 4×4 homogeneous transformation matrix which transforms a vector expressed in the j th coordinate system to the same vector expressed in the i th coordinate system

$(\cdot)^T$ transpose of $(\cdot)$

$\mathbf{J}_i$ 4×4 inertia matrix for link i

m_j mass of link j

$\overline{\mathbf{r}}_j$ 4×1 vector describing the center of mass with respect to the j th coordinate system

$\mathbf{g}$ 4×1 gravitational row vector, $[g_x, g_y, g_z, 0]$, and $|\mathbf{g}| = 9.8062\, m/s^2$

Appendix B
Recursive Newton-Euler Equations of Motion Referenced to the Link Coordinate Systems

Forward (or Outward) Equations: $i = 1, 2, \cdots, n$

$$ {}^i\mathbf{R}_0\boldsymbol{\omega}_i = {}^i\mathbf{R}_{i-1}[\, {}^{i-1}\mathbf{R}_0\boldsymbol{\omega}_{i-1} + \dot{q}_i\mathbf{z}_0\,(1-\lambda_i)\,] $$

$$ {}^i\mathbf{R}_0\dot{\boldsymbol{\omega}}_i = {}^i\mathbf{R}_{i-1}[\, {}^{i-1}\mathbf{R}_0\dot{\boldsymbol{\omega}}_{i-1} + (\ddot{q}_i\mathbf{z}_0 + {}^{i-1}\mathbf{R}_0\boldsymbol{\omega}_{i-1} \times \dot{q}_i\mathbf{z}_0\,(1-\lambda_i)\,] $$

$$ \begin{aligned} {}^i\mathbf{R}_0\dot{\mathbf{v}}_i = {} & ({}^i\mathbf{R}_0\dot{\boldsymbol{\omega}}_i) \times ({}^i\mathbf{R}_0\mathbf{p}_i^*) + ({}^i\mathbf{R}_0\boldsymbol{\omega}_i) \times [\,({}^i\mathbf{R}_0\boldsymbol{\omega}_i) \times ({}^i\mathbf{R}_0\mathbf{p}_i^*)\,] \\ & + {}^i\mathbf{R}_{i-1}(\ddot{q}_i\mathbf{z}_0\lambda_i + {}^{i-1}\mathbf{R}_0\dot{\mathbf{v}}_{i-1}) + 2\,({}^i\mathbf{R}_0\boldsymbol{\omega}_i) \times ({}^i\mathbf{R}_{i-1}\mathbf{z}_0\dot{q}_i)\lambda_i \end{aligned} $$

$$ {}^i\mathbf{R}_0\bar{\mathbf{a}}_i = ({}^i\mathbf{R}_0\dot{\boldsymbol{\omega}}_i) \times ({}^i\mathbf{R}_0\bar{\mathbf{s}}_i) + ({}^i\mathbf{R}_0\boldsymbol{\omega}_i) \times [({}^i\mathbf{R}_0\boldsymbol{\omega}_i) \times ({}^i\mathbf{R}_0\bar{\mathbf{s}}_i)] + {}^i\mathbf{R}_0\dot{\mathbf{v}}_i $$

Backward (or Inward) Equations: $i = n, n-1, \cdots, 1$

$$ {}^i\mathbf{R}_0\mathbf{f}_i = {}^i\mathbf{R}_{i+1}({}^{i+1}\mathbf{R}_0\mathbf{f}_{i+1}) + m_i\,{}^i\mathbf{R}_0\bar{\mathbf{a}}_i $$

$$ \begin{aligned} {}^i\mathbf{R}_0\mathbf{n}_i = {} & {}^i\mathbf{R}_{i+1}[\,{}^{i+1}\mathbf{R}_0\mathbf{n}_{i+1} + ({}^{i+1}\mathbf{R}_0\mathbf{p}_i^*) \times ({}^{i+1}\mathbf{R}_0\mathbf{f}_{i+1})] \\ & + ({}^i\mathbf{R}_0\mathbf{p}_i^* + {}^i\mathbf{R}_0\bar{\mathbf{s}}_i) \times (m_i\,{}^i\mathbf{R}_0\bar{\mathbf{a}}_i) \\ & + ({}^i\mathbf{R}_0\mathbf{I}_i\,{}^0\mathbf{R}_i)({}^i\mathbf{R}_0\dot{\boldsymbol{\omega}}_i) + ({}^i\mathbf{R}_0\boldsymbol{\omega}_i) \times [({}^i\mathbf{R}_0\mathbf{I}_i\,{}^0\mathbf{R}_i)({}^i\mathbf{R}_0\boldsymbol{\omega}_i)] \end{aligned} $$

$$ \tau_i = ({}^i\mathbf{R}_0\mathbf{n}_i)^T({}^i\mathbf{R}_{i-1}\mathbf{z}_0)(1-\lambda_i) + ({}^i\mathbf{R}_0\mathbf{f}_i)^T({}^i\mathbf{R}_{i-1}\mathbf{z}_0)\lambda_i + b_i\dot{q}_i $$

where $\mathbf{z}_0 = (0, 0, 1)^T$, $\lambda_i = 0$ if joint i is rotary, and $\lambda_i = 1$ if joint i is prismatic. The usual initial conditions are: $\boldsymbol{\omega}_0 = \dot{\boldsymbol{\omega}}_0 = \mathbf{v}_0 = \mathbf{0}$ and $\dot{\mathbf{v}}_0 = (g_x, g_y, g_z)^T$ (to include gravity), where $|\mathbf{g}| = 9.8062\ m/s^2$. Other variables are defined as:

b_i the viscous damping coefficient for joint i

$\boldsymbol{\omega}_i$ the angular velocity of link i with respect to the base coordinate system

$\dot{\boldsymbol{\omega}}_i$ the angular acceleration of link i with respect to the base coordinate system

$\mathbf{p}_i^*$ the origin of the i th frame with respect to the $i-1$ th frame

$\dot{\mathbf{v}}_i$ the linear acceleration of link i with respect to the base coordinate system

$\bar{\mathbf{a}}_i$ the linear acceleration of the center of mass of link i with respect to the base coordinate system

${}^{i-1}\mathbf{R}_i$ rotation matrix which maps position vectors from the i th coordinate system to the $i-1$ th coordinate system

$\mathbf{I}_i$ the inertia about center of mass of link i with respect to the i th coordinate system

$\mathbf{F}_i$ the total external force exerted on link i with respect to the base coordinate system

$\mathbf{N}_i$ the total moment exerted on link i with respect to the base coordinate system

$\mathbf{f}_i$ the force exerted on link i by link $i-1$ with respect to the base coordinate system ($\mathbf{f}_{n+1}$ = external force exerted on the hand)

$\mathbf{n}_i$ the moment exerted on link i by link $i-1$ with respect to the base coordinate system ($\mathbf{n}_{n+1}$ = external moment exerted on the hand)

5. REFERENCES

[1] M. E. Kahn and B. Roth, "The Near-Minimum-Time Control of Open-Loop Articulated Kinematic Chains," *Trans. of ASME, J. of Dynamic Systems, Measurement and Control,* Vol. 93, pp. 164-172, 1971.

[2] K. S. Fu, R. C. Gonzalez, and C. S. G. Lee, *Robotics: Control, Sensing, Vision, and Intelligence,* New York: McGraw-Hill, 580 pages, 1987.

[3] D. E. Orin, R. B. McGhee, M. Vukobratoric, and G. Hartoch, "Kinematic and Kinetic Analysis of Open-Chain Linkages Utilizing Newton-Euler Methods," *Math. Biosci.,* Vol. 43, pp. 107-130, 1979.

[4] J. Y. S. Luh, M. W. Walker, and R. P. C. Paul, "On-line Computational Scheme for Mechanical Manipulator," *Trans. ASME J. Dynam. Syst., Meas. Contr.,* Vol. 102, pp. 69-76, June 1980.

[5] D. E. Whitney, "The Mathematics of Coordinated Control of Prosthetic Arms and Manipulators," *Trans. of ASME, J. of Dynam. Syst., Meas. and Contr.,* Vol. 94, pp. 303-309, 1979.

[6] B. R. Markiewicz, "Analysis of the Computed Torque Drive Method and Comparison With Conventional Position Servo for a Computer Controlled Manipulator," Technical Memorandum 33-601, Jet Propulsion Laboratory, 1973.

[7] J. Y. S. Luh, M. W. Waker, and R. P. C. Paul, "Resolved-Acceleration Control of Mechanical Manipulators," *IEEE Trans. Automat. Contr.,* Vol. AC-25, No. 3, pp. 468-474, June 1980.

[8] E. G. Gilbert and I. J. Ha, "An Approach to Nonlinear Feedback Control with Application to Robotics," *IEEE Trans. Syst. Man, and Cybern.,* Vol. SMC-14, No. 6, pp. 879-884, Nov./Dec. 1984.

[9] T. J. Tarn et al., "Nonlinear Feedback in Robot Arm Control," *Proc. of 1984 Conf. on Decision and Control,* Las Vegas, Nevada, pp. 736-751, Dec. 1984.

[10] C. S. G. Lee and M. J. Chung, "An Adaptive Control Strategy for Mechanical Manipulators," *IEEE Trans. Automat. Contr.,* Vol. AC-29, No. 9, Sep. 1984, pp. 837-840.

[11] C. S. G. Lee and B. H. Lee, "Resolved Motion Adaptive Control for Mechanical Manipulators," *Trans. of ASME, J. of Dynam. Syst., Meas. and Contr.,* Vol. 106, No. 2, pp. 134-142, June 1984.

[12] J. J. Craig, P. Hsu, and S. S. Sastry, "Adaptive Control of Mechanical Manipulators," *Proc. of 1986 IEEE Int'l Conf. on Robotics and Automation,* pp. 190-195, Apr. 1986.

[13] J. J. E. Slotine and W. Li, "On the adaptive control of robot manipulators," *Int'l J. of Robotics Research,* Vol. 6 No. 3, pp. 49-59, Fall 1987.

[14] R. Johansson, "Adaptive control of robot manipulator motion," *IEEE Trans. Robotics and Automation,* Vol. 6, No. 4, pp. 483-490, Aug. 1990.

[15] R. P. Paul, "Modeling, Trajectory Calculation, and Servoing of a Computer Controlled Arm," Stanford Artificial Intelligence Laboratory Memo AM-177, Nov. 1972.

[16] A. K. Bejczy, "Robot Arm Dynamics and Control," Jet Propulsion Lab Technical Memo 33-669, Pasadena, CA., Feb. 1974.

[17] C. S. G. Lee, "Robot Arm Kinematics, Dynamics, and Control," *IEEE Computer,* Vol. 15, No. 12, pp. 62-80, Dec. 1982.

[18] D. E. Orin, "Pipelined Approach to Inverse Plant Plus Jacobian Control of Robot Manipulators," *Proc. of 1984 IEEE Int'l Conf. on Robotics and Automation,* pp. 169-175, Atlanta, GA, 1984.

[19] D. E. Orin, H. H. Chao and W. W. Schrader, "Pipeline/Parallel Algorithms for the Jacobian and Inverse Dynamics Computations," *Proc. 1985 IEEE Int'l Conf. on Robotics and Automation,* pp. 785-789, St. Louis, MO, 1985.

[20] R. P. Paul, *Robot Manipulators: Mathematics, Control, and Programming,* MIT Press, 1981.

[21] J. Y. S. Luh, "Conventional Controller Design for Industrial Robots - A Tutorial," *IEEE Trans. Syst. Man, and Cybern.,* Vol. SMC-13, No. 3, pp. 298-316, May/June 1983.

[22] C. S. G. Lee and P. R. Chang, "Efficient Parallel Algorithm for Robot Inverse Dynamics Computation," *IEEE Trans. Syst. Man, and Cybern.,* Vol. SMC-16, No. 4, pp. 532-542, July/Aug. 1986.

[23] C. S. G. Lee, T. N. Mudge, and J. L. Turney, "Hierarchical Control Structure Using Special Purpose Processors for the Control of Robot Arms," *Proc. of 1982 Pattern Recognition and Image Processing Conf.,* Las Vegas, Nevada, pp. 634-640, June 14-17, 1982.

[24] T. Kanade, P. K. Khosla, and N. Tanaka, "Real-Time Control of the CMU Direct Arm II Using Customized Inverse Dynamics," *Proc. of IEEE Conf. on Decision and Contr.,* pp. 1345-1352, Dec. 1984.

[25] R. Nigam, C. S. G. Lee, "A Multiprocessor-Based Controller for the Control of Mechanical Manipulators," *IEEE J. of Robotics and Automation,* Vol. RA-1, No. 4, pp. 173-182, Dec. 1985.

[26] E. Binder and J. Herzog, "Distributed Computer Architecture and Fast Parallel Algorithms in Real-time Robot Control," *IEEE Trans. Syst. Man, and Cybern.*, Vol. SMC-16, No. 4, pp. 543-549, July/Aug. 1986.

[27] L. H. Lathrop, "Parallelism in Manipulator Dynamics," *Int'l J. of Robotics Res.*, Vol. 4, No. 2, pp. 80-102, Summer 1985.

[28] A. Fijany and A. K. Bejczy, "A Class of Parallel Algorithms for Computation of the Manipulator Inertia Matrix," *IEEE Trans. Robotics and Automation,* Vol. RA-5, No. 5, pp. 600-615, Oct. 1989.

[29] J. Y. S. Luh and C. S. Lin, "Scheduling of Parallel Computer for a Computer-Controlled Mechanical Manipulator," *IEEE Trans. Syst. Man, and Cybern.*, Vol. SMC-12, No. 2, pp. 214-234, Mar. 1982.

[30] H. Kasahara and S. Narita, "Parallel Processing of Robot-Arm Control Computation on a Multiprocessor System," *IEEE J. of Robotics and Autom.*, Vol. RA-1, No. 2, pp. 104-113, June 1985.

[31] J. Barhen, "Robot Inverse Dynamics on a Concurrent Computation Ensemble," *Proc. of 1985 ASME Int'l Conf. on Computers in Engineering,* Vol. 3, pp. 415-429, 1985.

[32] M. Rahman and D. G. Meyer, "A Cost-efficient High-Performance Bit-serial Architecture for Robot Inverse Dynamics Computation," *IEEE Trans. Syst. Man, and Cybern.*, Vol. SMC-17, No. 6, pp. 1050-1058, Nov./Dec. 1987.

[33] C. L. Chen, C. S. G. Lee, and E. S. H. Hou, "Efficient Scheduling Algorithms of Robot Inverse Dynamics Computation on a Multiprocessor System," *IEEE Trans. Syst. Man, and Cybern.*, Vol. SMC-18, No. 5, pp. 729-743, Sep./Oct. 1988.

[34] C. S. G. Lee and C. L. Chen, "Efficient Mapping Algorithms for Scheduling Robot Inverse Dynamics Computation on a Multiprocessor System," *IEEE Trans. Syst. Man, and Cybern.*, Vol. SMC-20, No. 3, pp. 582-595, May/June 1990.

[35] J. M. Hollerbach, "A Recursive Lagrangian Formulation of Manipulator Dynamics and a Comparative Study of Dynamics Formulation Complexity," *IEEE Trans. Syst. Man, and Cybern.*, Vol. SMC-10, No. 11, pp. 730-736, Nov. 1980.

[36] P. M. Kogge, *The Architecture of Pipelined Computers,* McGraw Hill, New York, 1981.

[37] P. M. Kogge and H. S. Stone, "A Parallel Algorithm for the Efficient Solution of a General Class of Recurrence Equations," *IEEE Trans. Computers,* Vol. C-22, pp. 789-793, Aug. 1973.

[38] P. M. Kogge, "Parallel Solution of Recurrence Problems," *IBM J. Res. Develop.*, Vol. 18, pp. 138-148, Mar. 1974.

[39] E. Horowitz and S. Sahni, *Fundamentals of Computer Algorithms,* Computer Science Press Inc., pp. 488-494, 1978.

[40] D. A. Carlson, "Parallel Processing of Tree-like Computations," *Proc. of IEEE 4th Int'l Conf. on Distributed Computing System,* pp. 192-200, 1984.

[41] H. S. Stone, *Introduction to Computer Architecture,* Science Research Associate Inc., 1975.

[42] K. Hwang and F. A. Briggs, *Computer Architecture and Parallel Processing,* New York: McGraw Hill, 1984.

[43] B. Lint and T. Agerwala, "Communication Issues in the Design and Analysis of Parallel Algorithms," *IEEE Trans. Soft. Eng.,* Vol. SE-7, No. 2, Mar. 1981.

[44] T. Y. Feng, "A Survey of Interconnection Networks," *IEEE Computer,* pp. 12-27, Dec. 1981.

[45] C. L. Wu and T. Y. Feng, *Tutorial on Interconnection Networks for Parallel Distributed Processing,* IEEE Computer Society Press, 1984.

[46] H. S. Stone, "Parallel Processing with Perfect Shuffle," *IEEE Trans. Computers,* Vol. C-20, pp. 153-161, Feb. 1971.

[47] C. L. Wu, T. Y. Feng, "The Universality of the Shuffle-Exchange Network," *IEEE Trans. Computers,* Vol. C-30, No. 5, pp. 324-332, May 1981.

[48] C. Mead and L. Conway, *Introduction to VLSI Systems,* Addison-Wesley, 1980.

[49] H. T. Kung, R. Sproull, and G. Steele, *VLSI System and Computations,* Springer-Verlag, 1981.

[50] S. Y. Kung, K. S. Arun, J. Gal-Ezer, and D. V. Bhaskar Rao, "Wavefront Array Processor: Language, Architecture, and Applications," *IEEE Trans. Computers,* Vol. C-31, pp. 1054-1066, Nov. 1982.

[51] H. T. Kung, "Why Systolic Architectures?" *IEEE Computer*, pp. 37-46, Jan. 1982.

[52] C. S. G. Lee and P. R. Chang, "A Maximum Pipelined CORDIC Architecture for Robot Inverse Kinematic Position Computation," *IEEE J. of Robotics and Automation,* Vol. RA-3, No. 5, pp. 445-458, Oct. 1987.

[38] [illegible]

[illegible]

[illegible]

[illegible]

[42] K. Hwang and F. A. Briggs, *Computer Architecture and Parallel Processing*, New York: McGraw-Hill, 1984.

[43] B. Lint and T. Agerwala, "Communication Issues in the Design and Analysis of Parallel Algorithms," *IEEE Trans. Soft. Eng.*, Vol. SE-7, No. 2, Mar. 1981.

[44] T. Y. Feng, "A Survey of Interconnection Networks," *IEEE Computer*, pp. 12-27, Dec. 1981.

[45] C. L. Wu and T. Y. Feng, *Tutorial on Interconnection Networks for Parallel Distributed Processing*, IEEE Computer Society Press, 1984.

[46] H. S. Stone, "Parallel Processing with Perfect Shuffle," *IEEE Trans. Computers*, Vol. C-20, pp. 153-161, Feb. 1971.

[47] C. L. Wu and T. Y. Feng, "The Universality of the Shuffle-Exchange Network," *IEEE Trans. Computers*, Vol. C-30, No. 5, pp. 324-332, May 1981.

[48] C. Mead and L. Conway, *Introduction to VLSI Systems*, Addison-Wesley, 1980.

[49] H. T. Kung, R. Sproull, and G. Steele, *VLSI Systems and Computations*, Springer-Verlag, 1981.

[50] S. Y. Kung, K. S. Arun, R. J. Gal-Ezer, and D. V. Bhaskar Rao, "Wavefront Array [illegible]," *IEEE Trans. Computers*, Vol. C-31, pp. 1054-1066, Nov. 1982.

[51] H. T. Kung, "Why Systolic Architectures?" *IEEE Computer*, pp. 37-46, Jan. 1982.

[52] C. S. G. Lee and P. R. Chang, "A Maximum Pipelined CORDIC Architecture for Robot Inverse Kinematic Position Computation," *IEEE J. of Robotics and Automation*, Vol. RA-3, No. 5, pp. 445-458, Oct. 1987.

[illegible]

CHAPTER 2

NON-STRICT COMPUTATION OF THE MANIPULATOR INERTIA MATRIX

Masoud Amin-Javaheri
GMFanuc Robotics Corporation
2000 South Adams Road
Auburn Hills, MI 48326

David E. Orin
The Ohio State University
Department of Electrical Engineering
Columbus, Ohio 43210

I. Introduction

The real-time computational aspects of control schemes have been a problem under consideration over the years. In general, real-time computation is dependent upon two factors: state-of-the-art technological breakthroughs and overall task and system requirements. Since both of these factors are ever so rapidly changing, the implementation of computationally intensive algorithms must be continually re-evaluated to meet system requirements. In this chapter, the computational aspects of the manipulator inertia matrix is revisited with a new and effective computational approach which leads to a highly parallel algorithm suitable for multiprocessor architectural implementation.

Normally, to achieve speedup in computationally intensive algorithms, specifically, computation of the inertia matrix, parallel algorithms and architectures that adhere to the precedence relationships among processes have been the only avenues explored. That is, the precedence relationships among the processes dictate the flow of computation. This traditional computational approach is referred to here as a Strict Computational Model. The previously proposed parallel architectures for the inertia matrix [1] have been mostly pipelined architectures because of the inherent recurrence relationships within the algorithm. Throughput is increased substantially through pipelining; however, latency with use of a Strict Computational Model still remains as a major problem.

In addition, as the degree of parallelism within an algorithm that is based upon a Strict Computational Model is increased, the number of primitive operations needed increases correspondingly and the overall computation usually becomes much more involved. This rather general conclusion may be simply observed by comparing the $O(N)$ parallel algorithms presented in [1] with the $O(\log_2 N)$ parallel algorithms presented in [2] for the inertia matrix. Parallel algorithms based on a Strict Computational Model tend to escalate system complexity through the need for more processing elements, a more complex interconnection structure, and a higher communications bandwidth. All these factors adversely impact practical issues such as interconnection and system cost as well as more basic issues such as task scheduling, load balancing, and process synchronization.

To overcome the above problems, parallel algorithms in conjunction with a new computational model are devised to relax interprocess precedence for computing the inertia matrix which exhibits a high degree of process coupling. This high degree of process coupling is a result of the set of recurrence equations needed to compute the inertia matrix. Since computational precedence will not be strictly adhered to, the approach is referred to as Non-Strict Computation.

By relaxing the precedence among the processes in Non-Strict Computation of the inertia matrix, the critical path from input to output is shortened. In particular, since the inertia matrix is continually updated as the manipulator maneuvers in space, previous values computed are available and may be used as immediate inputs to processes to avoid delays based on strict precedence. The result is that the computational delay from input to output is effectively reduced. Also, with relaxation of precedence among processors, many of the problems mentioned above with escalating system complexity are significantly lessened.

Since strict precedence is not satisfied, the intermediate and final results of any Non-Strict Computation are imprecise. This problem may be remedied to a great extent by predicting the intermediate results passed among the processors. The degree of prediction will directly affect the level of accuracy in the results, and this will be explored when analyzing the Non-Strict Computational Approach.

Besides the data inaccuracy induced by the Non-Strict Computational Approach, there is still error in data values simply due to computational delay. However, since the computational delay caused by Non-Strict Computation is expected to be much lower than that due to Strict Computation, then the compound error in the data values may be less than that due to the sheer computational delay in the Strict Computation. An empirical measure of the overall computational delay has been developed in this work and is termed the effective latency. It represents the phase lag component of the total error as an equivalent delay of the computation and will be used to analyze the actual speedup realized with Non-Strict Computation.

While the computation of the inertia matrix is intensive, the inertia matrix itself is more slowly time-varying than some of the other terms within dynamics. Thus, its computation may be performed at a lower rate. This concept is referred to as multiple rate computation, and it will be applied to compute the acceleration-dependent term in the dynamic equations of motion given below:

$$\mathcal{T} = \mathbf{H}(\mathbf{q})\ddot{\mathbf{q}} + \mathbf{C}(\mathbf{q},\dot{\mathbf{q}}) + \mathbf{G}(\mathbf{q}) + \mathbf{J}^T(\mathbf{q})\mathbf{f} \tag{1}$$

where

$\mathcal{T}$	$N \times 1$	vector of joint torques (forces),
$\mathbf{q}, \dot{\mathbf{q}}, \ddot{\mathbf{q}}$	$N \times 1$	vectors of joint positions, velocities, and accelerations,
$\mathbf{C}(\mathbf{q},\dot{\mathbf{q}})$	$N \times 1$	vector of coriolis and centrifugal force terms,
$\mathbf{G}(\mathbf{q})$	$N \times 1$	vector of gravitational forces,
$\mathbf{J}(\mathbf{q})$	$6 \times N$	Jacobian matrix,
$\mathbf{f}$	6×1	vector of external forces and moments exerted by link N,
$\mathbf{H}(\mathbf{q})$	$N \times N$	symmetric, positive-definite, inertia matrix, and
N		the number of degrees of freedom of the manipulator.

In particular, in order to conduct a quantitative performance analysis, joint torques are computed by $\mathcal{T} = \mathbf{H}(\mathbf{q})\ddot{\mathbf{q}}$. Since the inertia matrix is slowly time-varying, its computation may be performed at a lower rate than the computation of the joint torques. The processors computing the inertia matrix will be idle and available for computing other tasks or terms such as the Coriolis terms. In so doing, a noticeable improvement in CPU availability is achieved. This major finding is demonstrated in a later section.

Parallel computation of the inertia matrix has also been considered in the context of computing robot forward dynamics [3] with a resulting algorithm which was of $O(N)$. More directly related, systolic architectures have been proposed in [1] to achieve an $O(N)$ computation time using N processors for an N degree-of-freedom manipulator. In [2], the development of an $O(\log_2 N)$ parallel algorithm for computing the manipulator inertia matrix is given. The latter two papers are based upon an efficient serial algorithm which uses the composite rigid body method [4]. In [5], similar algorithmic and architectural work based upon the composite rigid body spatial inertia is presented. While some amount of speedup is obtained with the various approaches taken in each of the above, it is usually realized at the expense of much increased system complexity. The alternative approach taken here based on Non-Strict Computation results in a relatively simple system architecture by relaxing the interprocess dependencies.

The Non-Strict Computational Approach was first introduced in [6] and applied to inverse dynamics analysis of a simple inverted pendulum. By relaxing precedence and thus reducing the critical path from input to output, significant speedup was achieved. However, the task flow graph for the inverse dynamics computation considered in [6] shows a high degree of parallelism among the processes. A multiprocessor implementation of it, then, based upon a Strict Computational Model, would result in a significant speedup, in many cases, even without relaxing precedence. On the other hand, the task flow graph for the inertia matrix given in [1] exhibits tightly-coupled sequential processes defined by a set of recurrence equations. This has placed tight precedence relationships among the processes, and as a result, the degree of parallelism achievable through a Strict Computational Model is limited. Hence, the Non-Strict Computational Model is even more significant when applied to the inertia matrix computation. Aslo, the use of prediction, not previously considered in [6], will allow the error to be reduced even further.

Linear extrapolation has been employed to obtain a parallel version of inverse kinematics [7]. A three-fold reduction in the computation time is achieved when the scheme is applied to the inverse kinematics solution of a PUMA 560. A similar scheme has also been applied to the inverse dynamics problem to decompose the Newton-Euler formulation [8] in order to increase parallelism. A mapping of the decoupled parallel recursive Newton-Euler algorithm for inverse dynamics onto a multiprocessor architecture is conducted in [9]. The goal in each of these is to speed up the computation while keeping the error within acceptable limits. While some sucess is achieved, it is helpful to directly compare the error resulting from both the Strict and Non-Strict approaches. The concept of an effective latency, as developed in this chapter, gives an important empirical measure to characterize the total error resulting from computational delay. The concept of Non-Strict Computation is developed and applied to the computation of the inertia matrix. Multiple rate computation and the effects of sampling frequency, not considered in the previous work above, are also explored in this chapter.

In the following, reference to three terms is constantly made: exact, strict, and non-strict computation. Exact computation refers to an ideal computation with no error and no delay; whereas, Strict Computation is the Exact Computation with computational delay included. On the other hand, Non-Strict Computation refers to a computation with generally less computational delay than the Strict Computation but with imprecise data.

In the next section, this chapter briefly describes the previous work completed in developing $O(N)$ parallel algorithms for the inertia matrix. Then, the Non-Strict Computational Approach in conjunction with the multiple rate concept is explained. Simulation results are then presented to show the effectiveness of this approach. Finally, the work is summarized and conclusions are made.

II. Parallel Algorithm for the Inertia Matrix

Previously, several parallel algorithms and architectures were devised to reduce the order of computation of $\mathbf{H}(\mathbf{q})$ from $O(N^2)$ to $O(N)$ [1]. With N processors configured in a linear array, the inertia matrix may be computed in row, column, or diagonal form. Fig. 1 shows the most efficient of these, which is the case of computing the inertia matrix in column form. In Fig. 1, θ_i (or more generally q_i) is joint angle i and

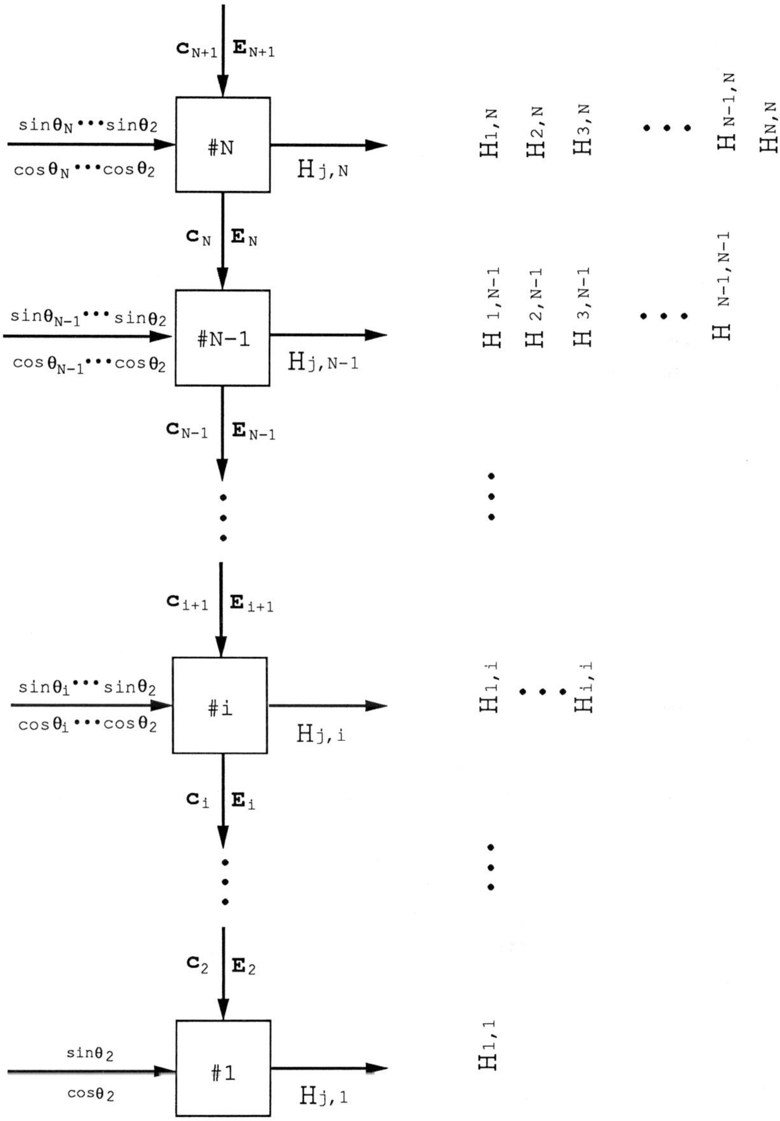

Figure 1: $O(N)$ Parallel Algorithm for Computing the Inertia Matrix in Column Form.

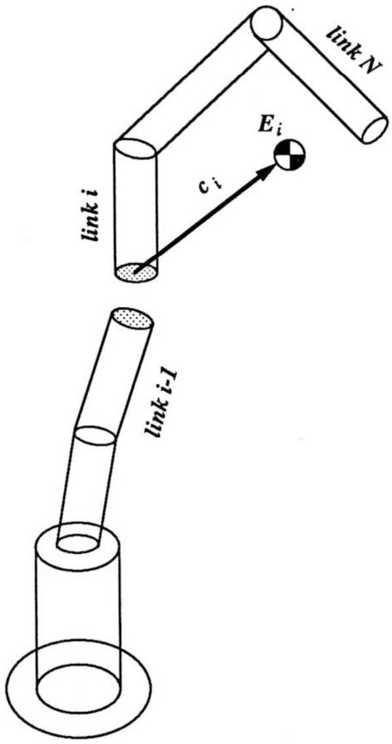

Figure 2: Two Composite Rigid Body Modeling of an Open-Chain Mechanism.

$H_{j,i}$ is the (j,i) element of the inertia matrix. Since the inertia matrix is symmetric, only the diagonal and upper off-diagonal elements of the inertia matrix are computed. For the column form, processor i computes the elements in the ith column of the matrix. The variables passed among processors are $\mathbf{c}_i$, the composite center of mass, and $\mathbf{E}_i$, the composite moment of inertia, for links i through N as shown in Fig. 2. The critical path for the computation involves the determination of $\mathbf{c}_i$ and $\mathbf{E}_i$ from $\mathbf{c}_{i+1}$ and $\mathbf{E}_{i+1}$. A linear recursive algorithm has been developed for this computation resulting in the $O(N)$ time for the inertia matrix. Also, the off-diagonal elements in each column of the inertia matrix are computed through a linear recursive algorithm beginning with the diagonal element and result in at most $O(N)$ time (for column $\#N$).

Fig. 3, a general timing diagram corresponding to Fig. 1, clearly indicates internal and external I/O plus the computational sequence needed to determine the inertia matrix. The darker line in the figure indicates the critical path which is from the input of the Nth processor to the output of the 1st processor. It is this data dependency among processors that has placed a lower bound on the compute time delay. It should be noted, though, that the use of distributed sampling by the individual processors

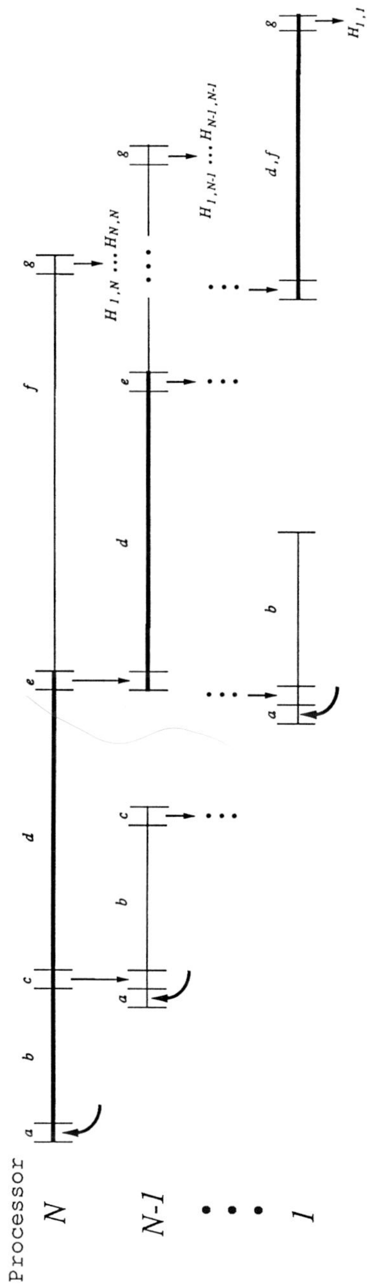

a : external I/O ($sin\theta's$, $cos\theta's$)
b : computation of $\mathbf{c}_i$
c : internal I/O ($\mathbf{c}_i$)
d : computation of $\mathbf{E}_i$
e : internal I/O ($\mathbf{E}_i$)
f : computation of $\mathbf{H}_{j,i}$
g : external I/O ($\mathbf{H}_{j,i}$)

Figure 3: Timing Diagram for Computing the Inertia Matrix.

on the input (sampling the input as needed) is assumed in the timing in Fig. 3. While this relaxes strict synchronization of input sampling, it has been shown to be effective in reducing the overall error [10].

The $O(N)$ parallel algorithm, which is shown in Fig. 3, maps onto a pipelined architecture to greatly improve the overall system throughput. However, such algorithms and architectures do not result in a significant reduction of compute time delay. This is generally true for algorithms mapped onto pipelined architectures. By taking a Non-Strict Computational Approach, which is the subject of the next section, we hope to significantly reduce the compute time delay by relaxing the data dependency while yet effectively managing the system complexity. Compute time delay is directly related to the phase margin and thus to the stability of a real-time control system. The lower the delay, the higher the margin of stability.

III. Non-Strict Computational Algorithm

As mentioned previously, the Non-Strict Computational Model is used to effectively relax precedence in computationally intensive tasks which need to be performed in real time. Relaxation of precedence leads to greater process parallelism and a shorter critical path from input to output. Consequently, a lower compute time delay is achieved. Relaxation of precedence implies that the data passed among processes are not precise. Imprecision in input data causes the resultant output data not to be exact either. However, the error caused by relaxing precedence does not accumulate in time. One way to improve the accuracy is to ensure that the intermediate results produced by the process nodes are carefully controlled to be close to the precise (or strict) values.

The use of prediction is one way to achieve the needed accuracy to ensure the integrity of the data. Simply, all or some of the variable values passed among nodes in a task graph may be predicted forward in time. Fig. 4 explicitly shows how predictors may be assigned to each node to relax precedence for the case of the column form shown in Fig. 1. In this case, the composite center of mass ($\mathbf{c}$) and the composite moment of inertia ($\mathbf{E}$) are predicted. Thus, processor i uses predicted values of ($\mathbf{c}_{i+1}$) and ($\mathbf{E}_{i+1}$) to compute ($\mathbf{c}_i$) and ($\mathbf{E}_i$) without waiting for the output of processor $i+1$. This simply implies that the arrows connecting processors indicate data flow but not precedence in computation. As a result, all N processors may carry on their corresponding computation in parallel. Thus, maximum process-level parallelism is achieved.

This point is clearly shown in the timing diagram in Fig. 5 corresponding to the Non-Strict Algorithm given in Fig. 4. Note that now the critical path is dominated by the computation of processor N only. Comparison of Figs. 3 and 5 reveals the fact that maximum process-level parallelism is achieved. This results in a speedup in the computation through a lower compute time delay from input to output. Thus, the effect of the Non-Strict Approach with prediction is to alter the critical path in

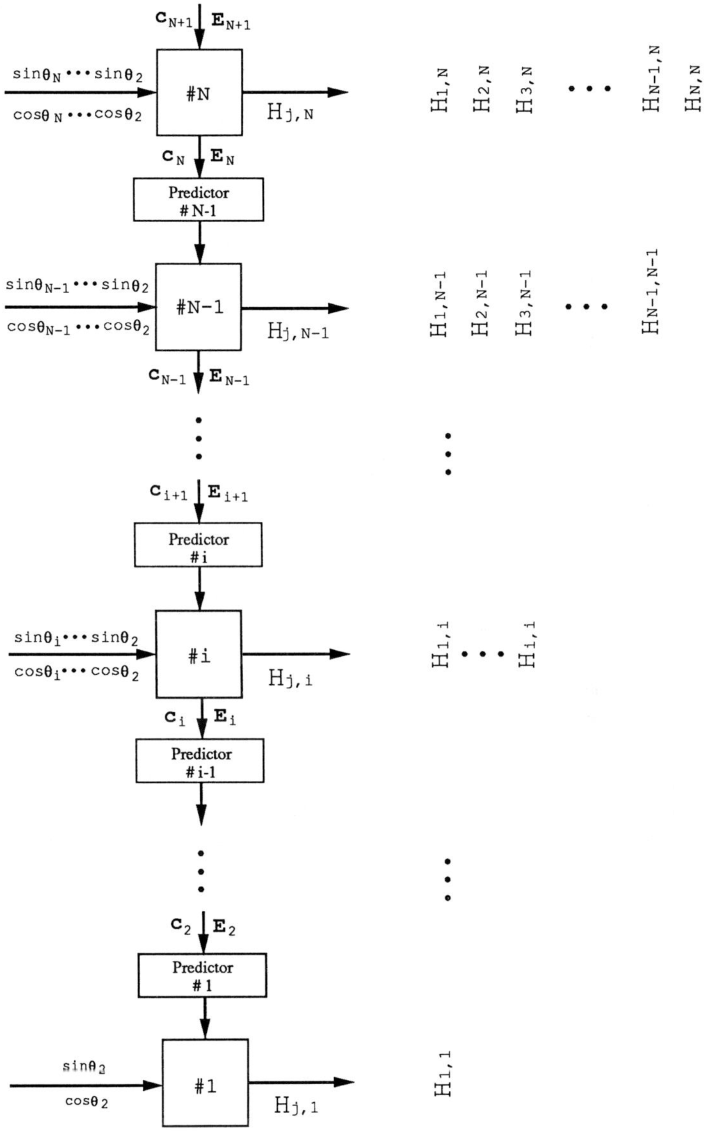

Figure 4: Non-Strict Computational Approach Showing the Use of Prediction.

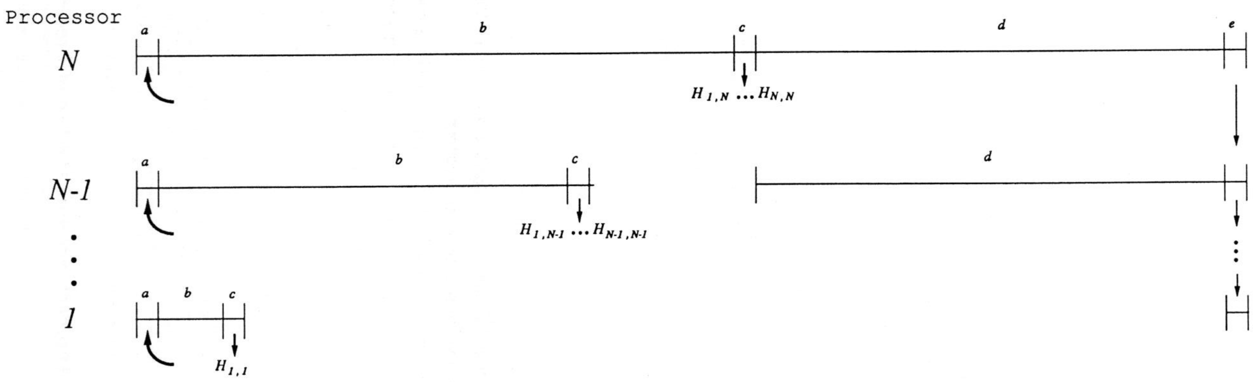

a : *external I/O (sinθ's , cosθ's)*

b : *computation of* $\boldsymbol{H}_{j,i}$

c : *external I/O* ($\boldsymbol{H}_{j,i}$)

d : *computation of* c_i & $\boldsymbol{E}_i$

e : *internal I/O* (c_i & $\boldsymbol{E}_i$)

Figure 5: Timing Diagram for Non-Strict Computation of the Inertia Matrix.

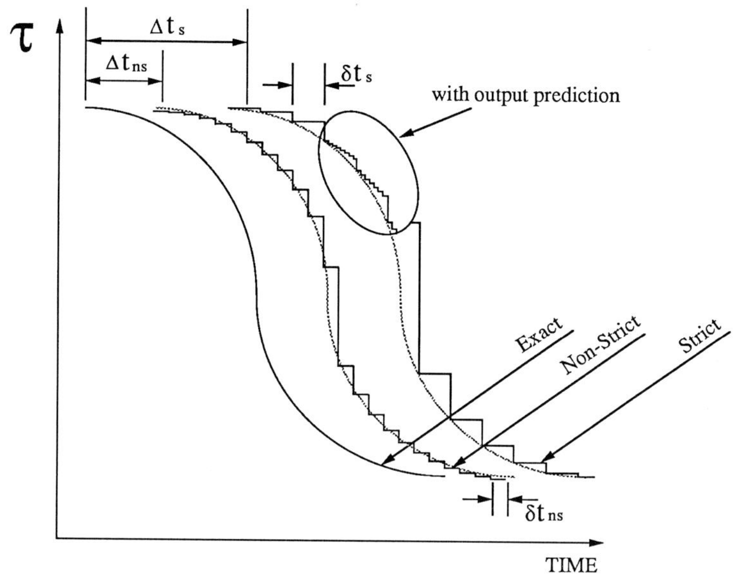

Figure 6: Various Computational Techniques.

order to lower the effective latency. As mentioned previously, lower effective latency results in a greater stability margin in a real-time control system.

Fig. 6 shows this reduction in the delay pictorially for a general time-dependent variable $\mathcal{T}$. Three distinct curves corresponding to exact, strict, and non-strict computations are shown. The delay, Δt_s, is the strict computational delay; whereas, Δt_{ns} is the non-strict computational delay. Note that, $\Delta t_{ns} < \Delta t_s$ if the non-strict approach is properly applied. In addition, since the critical path is shorter for the Non-Strict Computation than for the Strict Computation, the sampling rate is generally higher for the Non-Strict Computation. That is, $\delta t_{ns} < \delta t_s$.

Also, as shown in Fig. 6, the signal $\mathcal{T}$ is held constant during the entire sampling period for both the strict and non-strict curves. This sample and hold behavior gives rise to the "stair-step" response. The effective latency (T_{ed}) may then be approximated as the computational delay (T_c) plus one-half of the sampling period (T_s). That is,

$$T_{ed} \doteq T_c + \frac{1}{2}T_s, \tag{2}$$

where $T_{ed} = \Delta t_s$ for the Strict Computation or $T_{ed} = \Delta t_{ns}$ for the Non-Strict Computation. Clearly, the effective latency is a function of two independent variables: computational delay and sampling frequency. As discussed previously, the Non-Strict Computational Model may be employed to reduce computational delay and thus to reduce the effective latency. Also, suggested by Eq. (2), the higher the sampling frequency, the lower the effective latency would be.

Thus, Non-Strict Computation with prediction increases process-level parallelism. This directly impacts speedup through a lower effective latency, as mentioned previously. These desired enhancements are attainable at the expense of imprecise output results. From Fig. 6, note that the sampled points on the discrete non-strict curve do not necessarily lie on the continuous dotted curve, which is the exact curve delayed by the non-strict computational delay. Thus, the data are imprecise due to both delay and Non-Strict Computation. Prediction is used to reduce this delay. Further, by increasing the order of prediction, additional computational steps are needed, but the output error decreases. First-order prediction on $(\mathbf{c}_i)$ and $(\mathbf{E}_i)$ as opposed to zero-order prediction decreases the error noticeably at the expense of very minimal computational steps. However, higher orders of prediction may prove to be inefficient if the manipulator trajectory is fast-varying.

For zero-order prediction, simply the immediate previous value is used for the next sampling time. This is just the simplest form of extrapolation and gives:

$$H_{j,i}(t) = H_{j,i}(\mathbf{c}_i(t-1), \mathbf{E}_i(t-1)) \qquad \text{for} \quad i,j = 1, \ldots, N. \tag{3}$$

First-order prediction, on the other hand, uses the last two data points to linearly predict the current value of the function. This requires additional but minimal computational steps. The trade off between the computational accuracy and the extra computations incurred by the first-order predictor is well justified for the computation of the inertia matrix as will be demonstrated in Section 4.

It has already been mentioned that the effective latency may be reduced by lowering the compute time delay through Non-Strict Computation. Eq. (2) implies that to further decrease the effective latency substantially, the sampling frequency dependency should be eliminated or significantly diminished. By applying a multiple rate concept combined with output prediction to either the Strict or Non-Strict Computation, this may be accomplished. That is, as shown in Fig. 6 for the non-strict curve, by predicting $\mathcal{T}$ at a higher rate than the computation of the intermediate results, the effective latency would be primarily a function of computational delay, but not strongly a function of the sampling frequency. This idea is sketched in the circled portion of Fig. 6 by small step-size curves approaching the strict curve as the output rate increases or as the order of prediction on $\mathcal{T}$ increases. Thus, a maximum bound on the effective latency should be achievable at substantially lower sampling (computational) frequencies for the inertia matrix. A lower sampling frequency implies that CPU's are now available for other necessary computations. So, multiple rate computation in conjunction with prediction on the output should result in a higher CPU availability. In addition, the sampling frequency no longer plays as important a role in the determination of the effective latency.

As mentioned in the Introduction, the acceleration-dependent part of the joint torques will be computed ($\mathcal{T} = \mathbf{H}(\mathbf{q})\ddot{\mathbf{q}}$) for a quantitative performance analysis. The joint torques will be computed at a different rate than the computation of the inertia matrix – multiple rate computation of the joint torques. The resulting higher CPU availability gives time for other necessary computations to be computed along with the inertia matrix. The next section provides a number of quantitative results which demonstrate the above concepts.

IV. Simulation Results

In order to analyze the performance of the Non-Strict Computational Algorithms, the PUMA 560 robot manipulator is commanded to go through a smooth trajectory, based upon a fifth-order polynomial for joint position, at various acceleration rates and sampling frequencies. For typical applications of high-speed industrial robots, a smooth trajectory is essential and will only be considered in the analysis here.

At the outset, four sets of parameters are specified: the initial position and orientation, the goal position and orientation, sampling frequency, and total travel time (t_f). The initial position is arbitrarily chosen to be the zero position. The following Cartesian transform specifies the final position and orientation and causes the joint angles to sweep a wide range of the workspace.

$$\begin{bmatrix} 0 & -1 & 0 & -0.150\,m \\ 0 & 0 & 1 & 0.738\,m \\ -1 & 0 & 0 & -0.285\,m \\ 0 & 0 & 0 & 1 \end{bmatrix} . \tag{4}$$

All necessary dynamic parameters needed in the computation are derived from [11].

The position, velocity, and acceleration profiles for joint angle 1 are shown in Fig. 7 for $t_f = 4\,sec$. Other joints exhibit similar profiles at lower magnitudes. Thus, most of the discussion in this section is based upon joint 1 only. It should be pointed out that a $4\,sec$ total travel time is near the fastest that the PUMA 560 is able to travel without exceeding the joint torque limits for the initial and final positions as chosen.

As mentioned before, the joint torques are computed using the dynamic equations of motion, but using only the acceleration-dependent term affected by the inertia matrix. For joint 1, the torque computed by $\mathcal{T} = \mathbf{H}(\mathbf{q})\ddot{\mathbf{q}}$ is shown in Fig. 8 (exact case with no delay).

Based upon the results of [1] and for purposes of illustration here, the computation time for each processor in the Non-Strict case (Fig. 5) is set to $1\,msec$. For Strict Computation of the inertia matrix for a six degree of freedom manipulator ($N = 6$), the delay along the critical path (Fig. 3) would be $1.75\,msec$ [1]. Thus, for Strict Computation of the manipulator inertia matrix, variation of the joint torques is the

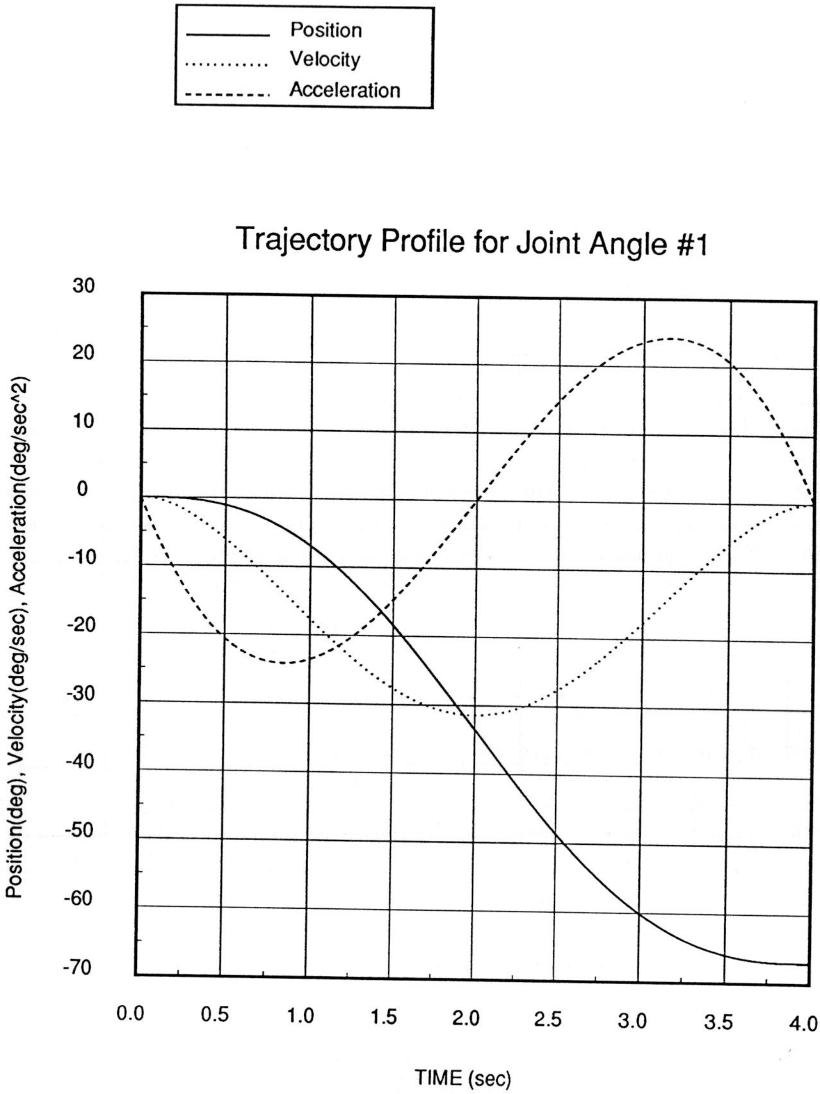

Figure 7: Trajectory Profile for Joint Angle #1 ($t_f = 4\ sec$).

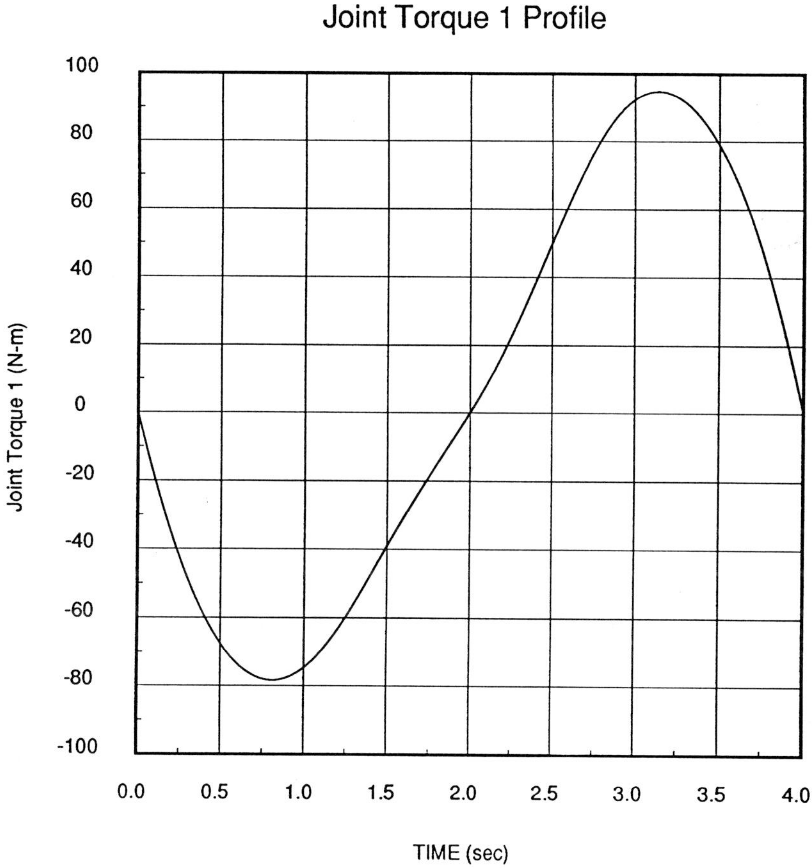

Figure 8: Exact Computation of Joint Torque 1.

same as the exact case except that all the curves are delayed by $1.75\,msec$. Based upon a $1.0\,msec$ computation time for the Non-Strict Computation, the $(a+b+c)$ computation (Fig. 5) is approximately 63% of the overall computation [1]. Thus, the effective latency for the non-strict case can be as small as $0.63\,msec$. The maximum sampling frequency for the Strict Computation is $\frac{1}{1.75\,msec} = 570\,\mathrm{Hz}$; whereas, the maximum sampling frequency for the Non-Strict Computation is $\frac{1}{1\,msec} = 1\,\mathrm{KHz}$. The reason behind achieving a higher sampling frequency (from 570 to 1 KHz) is due to increased parallelism as mentioned previously. Thus, a higher sampling rate and a lower effective latency are obtained.

A pipelined architecture implementing the $O(N)$ parallel algorithm shown in Fig. 1 would result in a sampling rate close to 1 KHz and a corresponding effective delay of about $1msec$. The effective latency is reduced by about 70% (from $1.75msec$ to $0.63msec$) for the Non-Strict Computation; whereas, it is only reduced by about 40% (from $1.75msec$ to about $1msec$) for the pipelined architecture for a six degree-of-freedom manipulator [1]. An analysis of the effective latency for the Non-Strict Computation is comprehensively analyzed in the remainder of this section.

A. Non-Strict Results with Zero-Order Output Prediction

In order to first investigate the effects of sampling frequency and multiple rate computation, the simulator is used to compute joint torques at various sampling frequencies. The effective latency is computed by an incremental time advancement of the exact data points until a best fit is obtained with the non-strict data points. The standard deviation is used to evaluate the best fit. While searching for the best fit between the exact and non-strict data points, there is a point in time when standard deviation reaches its minimum value. This time is reported as the effective latency. Mathematically,

$$\Delta t_{ns} \;:\; \min\left\{(\delta\mathcal{T})^2 = \int_{\Delta}^{t_f-\Delta}\left(\mathcal{T}^{exact}(t+\Delta t_{ns}) - \mathcal{T}^{non-strict}(t)\right)^2 dt\right\} \tag{5}$$

where Δ is the initial (startup) delay due to recursion in the non-strict computation and varies as a function of joint number. The above definition for obtaining the effective latency is applied to each joint separately.

Table 1 shows the effective latencies for only the major axes for various sampling frequencies. For every joint, three rows of data are given. The first row is the effective latency using Strict Computation while the second and third rows correspond to zero-order and first-order prediction on internal I/O variables passed among processors, respectively. The time increment used in computing the effective latency is $10\,\mu sec$; so, the resolution for the data points in Table 1 is $\pm 10\,\mu sec$.

From Table 1, it may be observed that for a given joint, the effective latency, in general, decreases in a given column for each joint. That is, Non-Strict Computation offers a lower effective latency than the Strict Computation, as is expected. For

Table 1: Effective Latency with Zero-Order Output Prediction ($t_f = 4\,sec$). (First row of three – Strict, second row – Zero-Order Non-Strict, third row – First-Order Non-Strict)

Joints	Effective Latency ($msec$)				
	$f_s = 72.5$ Hz	$f_s = 145$ Hz	$f_s = 285$ Hz	$f_s = 570$ Hz	$f_s = 1$ KHz
1	8.66	5.19	3.47	2.62	-
	9.11	4.81	2.71	1.68	1.22
	7.52	4.07	2.36	1.51	1.12
2	8.67	5.19	3.47	2.62	-
	7.55	4.09	2.36	1.51	1.13
	7.54	4.08	2.36	1.51	1.13
3	8.68	5.21	3.48	2.62	-
	7.53	4.09	2.36	1.51	1.14
	7.54	4.09	2.36	1.51	1.14

example, there is about a 42% reduction in the effective latency for joint angle 1 at 570 Hz, from 2.62 $msec$ to 1.51 $msec$ corresponding to first-order prediction. Note that the major reduction in the effective latency, about 36%, is due to Non-Strict Computation (from 2.62 $msec$ to 1.68 $msec$) and only the remaining 6% reduction in effective latency is due to the order of prediction (from 1.68 $msec$ to 1.51 $msec$). Thus for this case, the order of prediction does not have a major impact on the effective latency. A more reasonable comparison is the effective latency between the Strict Computation at its highest frequency (570 Hz) and the Non-Strict Computation (first-order prediction) at its highest frequency (1 KHz). In this case, there is about a 57% reduction in the effective latency, from 2.62 $msec$ to 1.12 $msec$. This reduction holds for all joints and corresponds to a speedup factor of 2.3.

By scanning a given row for any of the joints, it is obvious that the effective latency decreases substantially as the sampling frequency increases for either Strict or Non-Strict Computation. There is about a 70% reduction in the effective delay for joint 1 by increasing the sampling frequency from 72.5 Hz to 570 Hz in the Strict Computation; whereas, there is an approximate 85% reduction in the effective latency by increasing the sampling frequency from 72.5 Hz to 1 KHz for the Non-Strict Computation (zero-order or first-order) for the same joint.

The above findings have re-emphasized two major results: the effectiveness of the Non-Strict Computation in reducing the effective latency and more importantly the dependency of the effective latency on the sampling frequency. Close observation of the numbers in Table 1 confirm the validity of Eq. (2). For example, at $T_s = \frac{1}{f_s} = \frac{1}{570} = 1.75\,msec$, the effective latency is $0.87\,msec$ (one-half sampling period) plus $1.75\,msec$ (strict computational delay) which is equal to $2.62\,msec$, and is equal to the number obtained through simulation. For the Non-Strict Computation, the computational delay is only $0.63\,msec$ as opposed to $1.75\,msec$. Consequently, for the Non-Strict Computation, the sampling frequency has a higher weighting factor than the computational delay toward the determination of the effective latency.

B. Non-Strict Results with Multiple Rate Computation

As was proposed in a previous section, multiple rate computation plus output prediction may be incorporated to eliminate or to diminish the effect of the sampling frequency. Results similar to those given in Table 1 are tabulated in Table 2 in which first-order multiple rate, output prediction on joint torques is applied. Now, for joint angle 1, there is a 63% reduction in the effective latency, from $1.75\,msec$ at $f_s = 570$ Hz to $0.64\,msec$ at $f_s = 1$ KHz. This corresponds to a factor of 2.7 speedup. Also, comparing the two tables shows a much lower effective latency at a given frequency for all the joints. In Table 2, the effective latency is more greatly influenced by computational delay and less by the sampling frequency. This result is very pronounced at high frequencies. By scanning the rows for a given joint, it is clear that the effective latency approaches its minimum bound, the computational delay which is equal to $0.63\,msec$, at much lower frequencies. For example, the third row for any joint indicates that the effective latency converges to the computational delay at frequencies as low as 72.5 Hz. This frequency corresponds to a $13.8\,msec$ sampling period as compared to a $1.0\,msec$ time needed to compute the manipulator inertia matrix. In other words, by using a sampling frequency of 72.5 Hz, every processor is now available $12.8\,msec$ of the time. This implies a 93% CPU availability.

By employing multiple rate computation with output prediction, the following three major results are attainable:

1. A good speedup factor,
2. A low effective latency which is more strongly related to the computational delay and less dependent upon the sampling frequency, and
3. A high CPU availability for low sampling frequencies.

Table 2: Effective Latency with First-Order, Multiple Rate Output Prediction ($t_f = 4\,sec$). (First row of three – Strict, second row – Zero-Order Non-Strict, third row – First-Order Non-Strict)

	Effective Latency ($msec$)				
Joints	$f_s = 72.5\,\text{Hz}$	$f_s = 145\,\text{Hz}$	$f_s = 285\,\text{Hz}$	$f_s = 570\,\text{Hz}$	$f_s = 1\,\text{KHz}$
	1.76	1.75	1.75	1.75	-
1	2.26	1.39	1.00	0.82	0.74
	0.65	0.64	0.64	0.64	0.64
	1.77	1.75	1.75	1.75	-
2	0.69	0.65	0.65	0.64	0.64
	0.67	0.65	0.64	0.64	0.64
	1.73	1.74	1.75	1.75	-
3	0.63	0.64	0.64	0.64	0.64
	0.63	0.64	0.64	0.64	0.64

C. Error Analysis

The only drawback of Non-Strict Computation is the output error due to data imprecision. This effect is indirectly displayed in Tables 1 and 2. For example in Table 1, for the Non-Strict Computation at 1 KHz with zero-order output prediction, the effective latency is about $0.5\,msec$ (one-half sampling period) plus $0.63\,msec$ (non-strict computational delay) which is equal to $1.13\,msec$ as compared $1.22 \pm 0.01\,msec$ for zero-order prediction and $1.12 \pm 0.01\,msec$ for first-order prediction. The major discrepancy between 1.13 and $1.22\,msec$ is the error in the data due to the non-strict nature of the computation. But, note that $1.13\,msec$ is very close to $1.12\,msec$. This simply emphasizes the fact that higher-order prediction (first-order in this case) may be an effective technique to reduce the error. This result may also be revealed from Table 2 by simply comparing the second and the third row for joint angle 1. For instance, at a 72.5 Hz sampling frequency, the effective latency is $2.26\,msec$ when zero-order prediction is employed; whereas, this number approaches its minimum value of $0.63\,msec$ when first-order prediction is employed.

To more directly show the effect of first-order prediction on the joint torques computed using $\mathcal{T} = \mathbf{H}(\mathbf{q})\ddot{\mathbf{q}}$, the following performance curves are computed.

$$\begin{aligned} \mathcal{T}^{es} &= \mathcal{T}^{exact} - \mathcal{T}^{strict}, \\ \mathcal{T}^{en} &= \mathcal{T}^{exact} - \mathcal{T}^{non-strict}, \\ \mathcal{T}^{bn} &= \mathcal{T}^{bestfit} - \mathcal{T}^{non-strict}, \end{aligned} \tag{6}$$

where non-strict corresponds to both zero-order and first-order prediction. Overall, this corresponds to five different data sets which are graphically shown in Fig. 9 for joint #1 only. The percentage of the peak-to-peak variation of all five curves is used as a performance measure of output error due to delay and/or data corruption. Thus, comparison of $\mathcal{T}^{en}$ and $\mathcal{T}^{bn}$ for zero-order prediction based upon this performance measure reveals that less than 20% of the error is due to non-strict data corruption. On the other hand, comparison of $\mathcal{T}^{en}$ and $\mathcal{T}^{bn}$ for first-order prediction indicates that approximately 98% of the error is due to delay and only 2% of the error is due to non-strict data corruption. This simply demonstrates that first-order prediction significantly reduces the error. Comparison of the two $\mathcal{T}^{bn}$ curves for zero-order and first-order prediction is another way to restate the above finding. Simply put, a 90% improvement in error is achieved when first-order prediction is employed. Comparison of $\mathcal{T}^{es}$ and $\mathcal{T}^{bn}$ for first-order prediction indicates a 98% overall improvement in error due to either delay or data corruption. Finally, the torque errors at higher frequencies, for example 1 KHz which are not explicitly shown, are generally lower by an order of magnitude.

The results obtained above indicate that a combination of first-order output prediction and first-order prediction on intermediate result values passed between processors result in the best overall response. First-order output prediction allows a reduction in the acceptable sampling frequency (72.5 Hz instead of 1 KHz), increased CPU availability (by about 93%), and a 2.7 speedup factor (i.e. lower effective latency); whereas, first-order prediction on intermediate result values reduces the error created by Non-Strict Computation to a negligible amount.

D. Effect of Trajectory Profile

The effective latency illustrated in Tables 1 and 2 varies as the trajectory changes and, of course, as the acceleration changes. To demonstrate the latter effect, the total travel time is changed. By increasing the acceleration, no change in speedup is expected. But, the sampling period significantly impacts the effective latency. Under the exact operating conditions as before, with the exception of $t_f = 0.4\,sec$, a $0.64\,msec$ effective latency is achieved at an approximate 290 Hz sampling frequency instead of the 72.5 Hz obtained at $t_f = 4\,sec$. This corresponds to a 71% CPU availability which is still substantial. In addition, reducing the total travel time by a factor of 10, which is not realizable with the PUMA 560, increases the absolute

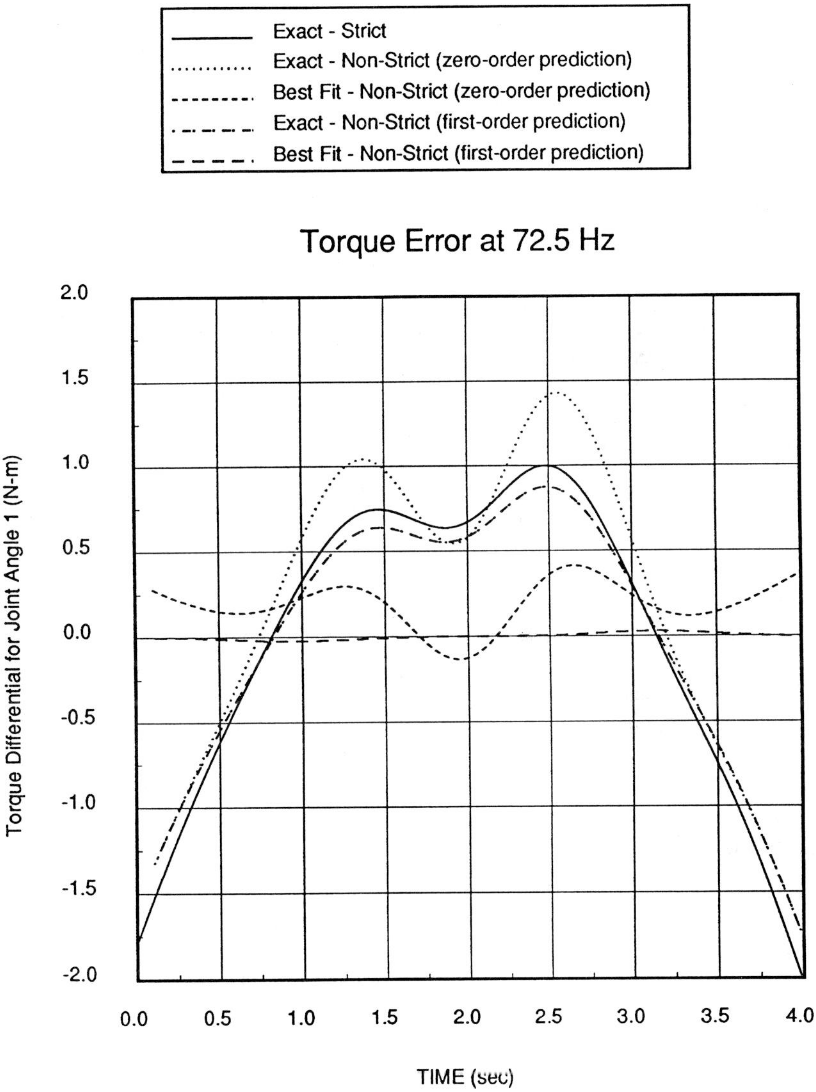

Figure 9: Joint Torque Error for Joint Torque #1.

differential torque error by a factor of 1000. Nonetheless, similar improvements in the percentage of error is attainable when first-order prediction is applied.

When increasing the total travel time and thus decreasing accelerations, the effective latency approaches the computational delay at frequencies lower than 72 Hz. This simply implies even higher CPU availabilities. Also, the magnitude of the differential torque error decreases correspondingly. Thus, acceleration profile has a direct impact on the effective latency and CPU availability.

V. Summary and Conclusions

This chapter has outlined the development of a highly parallel algorithm based on the Non-Strict Computational Approach for the manipulator inertia matrix. The computation of the inertia matrix is recursive in nature with a strong degree of coupling among processes. However, a high degree of parallelism can be achieved by relaxing precedence. Consequently, a shorter critical path from input to output resulting in a lower effective latency is obtained at the expense of imprecise data. In order to ensure the integrity of the results, prediction has been applied along with the relaxation of precedence so as to control the error between the non-strict and strict values. In addition, increasing the order of prediction may lower the severity of data error at the expense of extra but minimal computational steps when the proposed algorithm is actually mapped onto a suitable parallel architecture.

Multiple rate computation combined with output prediction has also been applied to determine the acceleration-dependent term of the dynamic equations of motion. The computation of the inertia matrix was performed at lower rates than the rate of computation of the joint torques, with highly improved performance for the case of Non-Strict Computation. In so doing, the effect of sampling rate, which greatly influences the effective latencies as given by the approximation in Eq. (2), was drastically diminished for both the strict and non-strict cases. This led to an approximate 90% CPU availability during which the parallel architecture implementing the inertia matrix is free to compute other tasks.

All the foregoing concepts have been demonstrated by simulating a multiprocessor architecture computing the acceleration-dependent part of the dynamic equations of motion for the PUMA 560. The effective latency is computed for both the Strict and Non-Strict (for both zero-order and first-order prediction) Computations at various sampling frequencies as shown in Table 1. Table 2 presents similar data when multiple rate computation combined with first-order prediction on the output is used. Comparison of the data in the two tables simply reflects the fact that sampling frequency has a negligible effect on the effective latency when joint torques are computed at a higher rate, with first-order output prediction, than the computation of the inertia matrix. Also, a reasonable effective latency is achievable at much lower sampling frequencies leading to a high CPU availability.

The joint torque errors due to both computational delay and Non-Strict Computation for joint #1, as a typical case, are shown in Fig. 9. Comparison of the curves in the figure indicates that the error may be noticeably reduced if the order of prediction is increased. Finally, the effect of acceleration and total travel time along the trajectory have been discussed.

This chapter, in general, has presented a practical case study and the foundation for the Non-Strict Computational Approach in highly coupled algorithms. Further applications of the Non-Strict Computational Approach need to be explored. Then, a formalization of the approach may be developed for a general class of computationally intensive problems. Hopefully, this will result in a wide applicability of the approach giving significant speedups while effectively managing parallel system complexity.

REFERENCES

1. M. Amin-Javaheri and D. E. Orin, "Systolic architectures for the manipulator inertia matrix," *IEEE Transactions on Systems, Man, and Cybernetics*, vol. 18, no. 6, pp. 939–951, November/December 1988.

2. M. Amin-Javaheri and D. E. Orin, "Parallel algorithms for computation of the manipulator inertia matrix," to appear in *International Journal of Robotics Research, MIT Press*, 1991.

3. C. S. G. Lee and P. R. Chang, "Efficient parallel algorithms for robot forward dynamics computation," *IEEE Transactions on Systems, Man, and Cybernetics*, vol. SMC-18, no. 2, pp. 238–251, March/April 1988.

4. M. W. Walker and D. E. Orin, "Efficient dynamic computer simulation of robotic mechanisms," *Journal of Dynamic Systems, Measurement, and Control*, vol. 104, pp. 205–211, September 1982.

5. A. Fijany and A. K. Bejczy, "A class of parallel algorithms for computation of the manipulator inertia matrix," in *Proc. of IEEE International Conference on Robotics and Automation*, pp. 1818–1826, Scottsdale, Arizona, May 1989.

6. D. E. Orin, P. Sadayappan, and H. Wong, "Non-Strict computational approach: example of the dynamics of a simple inverted pendulum," *OSU Internal Publication*, June 1988.

7. H. Zhang and R. P. Paul, "A parallel inverse kinematics solution for robot manipulators based on multiprocessing and linear extrapolation," in *Proc. of IEEE International Conference on Robotics and Automation*, pp. 468–474, Cincinnati, Ohio, May 1990.

8. E. E. Binder and J. H. Herzog, "Distributed computer architecture and fast parallel algorithms in real-time robot control," *IEEE Transactions on Systems, Man, and Cybernetics*, vol. SMC-16, no. 4, pp. 543–549, July/August 1986.

9. M. Vuskovic, T. Liang, and K. Anantha, "Decoupled parallel recursive Newton-Euler algorithm for Inverse Dynamics," in *Proc. of IEEE International Conference on Robotics and Automation*, pp. 832–838, Cincinnati, Ohio, May 1990.

10. J. N. Anderson and A. Smiarowski, "A specialized parallel architecture for robotic controllers," in *Second International Symposium on Robotics and Manufacturing*, Albuquerque, NM, November 1988.

11. B. Armstrong, O. Khatib, and J. Burdick, "The explicit dynamic model and inertial parameters of the PUMA 560 arm," in *Proc. of IEEE International Conference on Robotics and Automation*, pp. 510–518, San Francisco, CA, 1986.

CHAPTER 3

PARALLEL PROCESSING OF ROBOT CONTROL AND SIMULATION

Hironori Kasahara
Department of Electrical Engineering, Waseda University
3-4-1 Ohkubo Shinjuku-ku, Tokyo, 169, Japan

ABSTRACT

This chapter describes a parallel processing scheme for robot- arm dynamic control computation and robot-arm dynamics simulation. Parallel processing of robot control computation and simulation has attracted much attention to develop cost- effective and advanced controllers and simulators. The parallel processing scheme allows us to solve the inverse dynamic problem for control and the forward dynamics problem for simulation in the minimum time on a multiprocessor system with an arbitrary number of processors. The scheme consists of the following steps: generation of relatively coarse grain tasks each of which includes multiple operations of small matrices and vectors, analysis of data dependences among the tasks, scheduling of the tasks to processors and generation of optimized parallel machine codes.

1. INTRODUCTION

Along with the introduction of robot systems into CIM (Computer Integrated Manufacturing) and hazardous jobs in space, ocean floor, and radio active pollution areas, the demand for more advanced robot systems which can quickly and accurately perform complicated operations has been increasing. In the development of such robot systems, it is very important to put advanced control schemes like dynamic control to practical use and to develop real-time dynamics simulators for evaluation of the control schemes and robot configuration. However, the dynamic control computation [12][14][15], which is the computation to obtain force or torque applied to each robot joint for a desirable motion trajectory, and the dynamics simulation [13], which calculates robot motion (joint displacement, velocity and acceleration) for given force or torque applied to each joint, contain a heavy computational load composed of a lot of operations of small matrices and vectors. For this reason, real-time processing of control and simulation calculation requires an expensive mini- or supermini-computer having high processing capability. In the light of this fact, recently, much attention has been paid to parallel processing using a plurality of inexpensive microprocessors in order to improve the processing capability and cost performance of a robot controller or simulators[1]-[10].

To develop a parallel processing system for the robot-arm control and simulation, we can use several types of parallel processing techniques such as vector pipeline processors being used in current supercomputers, data-flow machines and multiprocessor systems. Vector pipeline processors, however, can not efficiently handle operations for small matrices, or short vectors, since overhead caused by an initial pipe delay is relatively large. The data-flow machines also may have difficulties to deterministically manage a sampling period for on-line real-time computation since data driven execution of the data-flow machines is generally non-deterministic. Therefore, the multiprocessor systems [16] appear to be most suitable for real-time robot control and simulation.

Many parallel processing schemes of the robot-arm dynamic control and dynamics simulation based on the Newton-Euler equations[3][7][8], which is the most efficient sequential algorithm to solve the inverse dynamic problem, for multiprocessor systems have so far been proposed. Those schemes can be divided into two different approaches. The former is a hardware oriented approach that designs such a dedicated architecture as fits a computational flow of the control and simulation computation. The latter is a software oriented approach that develops a special purpose compiler which decomposes the control and simulation computation into tasks, schedules the tasks to processors and generates parallel machine codes for an arbitrary architecture of multiprocessor system [1][2][4].

In this chapter, a software oriented parallel processing scheme of the robot-arm dynamic control using Newton-Euler equations [1][12] and the dynamics simulation using Walker and Orin's algorithm [13] based on the Newton-Euler equations is described. Furthermore, its practicality and effectiveness are shown on an actual multimicroprocessor system.

2. ROBOT-ARM DYNAMIC CONTROL COMPUTATION AND DYNAMICS SIMULATION

This section briefly describes features of the Newton-Euler equations for the robot-arm control computation and the Walker & Orin's algorithm for the simulation.

2.1 Robot-arm control computation

Robot-arm control computation generally involves a lot of operations for small size matrices and vectors related with coordinate transformations. For example, the Newton-Euler equations for motion consists of many 3 by 3 matrix operations and 3 by 1 vector operations as shown in Fig. 1. It has been known that Newton-Euler equations has the lowest computational complexity, or **O**(n) where **O** is the order function and n is a number of robot joints. In this figure, the subscript i stands for the i-th joint of the arm. The equations above the dashed line are calculated repeatedly for i = 1 to n and those below the line for i = n to 1 in the reverse direction. In these equations, variables A_i^{i+1} and J_i 3 by 3 matrices, $\dot{q}_i$, $\ddot{q}_i$, m_i, b_i, and t_i are scalars and the remaining variables are 3 by 1 vectors. Parallel processing of the Newton-Euler equations has been studied by many researchers since Luh proposed the parallel processing for the first time[1].

$$A_i^0\omega_i=\begin{cases} A_i^{i-1}(A_{i-1}^0\omega_{i-1}+z_0\dot{q}_i), \text{ if link } i \text{ is rotational,} \\ A_i^{i-1}(A_{i-1}^0\omega_{i-1}), \text{ if link } i \text{ is translational,} \end{cases}$$

$$A_i^0\dot{\omega}_i=\begin{cases} A_i^{i-1}[A_{i-1}^0\dot{\omega}_{i-1}+z_0\ddot{q}_i+(A_{i-1}^0\omega_{i-1})\times(z_0\dot{q}_i)], \\ \quad \text{if link } i \text{ is rotational,} \\ A_i^{i-1}(A_{i-1}^0\dot{\omega}_{i-1}), \text{ if link } i \text{ is translational,} \end{cases}$$

$$A_i^0\dot{\nu}_i=\begin{cases} (A_i^0\dot{\omega}_i)\times(A_i^0p_i^*)+(A_i^0\omega_i)\times[(A_i^0\omega_i)\times(A_i^0p_i^*)]+A_i^{i-1}(A_{i-1}^0\dot{\nu}_{i-1}), \\ \quad \text{if link } i \text{ is rotational,} \\ A_i^{i-1}(z_0\ddot{q}_i+A_{i-1}^0\dot{\nu}_{i-1})+(A_{i-1}^0\dot{\omega}_i)\times(A_i^0p_i^*) \\ \quad +2(A_i^0\omega_i)\times(A_i^{i-1}z_0\dot{q}_i)+(A_i^0\omega_i)\times[(A_i^0\omega_i)\times(A_i^0p_i^*)], \\ \quad \text{if link } i \text{ is translational,} \end{cases}$$

$$A_i^0\dot{\hat{v}}_i=(A_i^0\dot{\omega})\times(A_i^0\hat{s}_i)+(A_i^0\omega_i)\times[(A_i^0\omega_i)\times(A_i^0\hat{s}_i)]+A_i^0\dot{\nu}_i$$

$$A_i^0F_i=m_iA_i^0\dot{\hat{v}}_i$$

$$A_i^0N_i=(A_i^0J_iA_0^i)(A_i^0\dot{\omega}_i)+(A_i^0\omega_i)\times[(A_i^0J_iA_0^i)(A_i^0\omega_i)]$$

$$A_i^0f_i=A_i^{i+1}(A_{i+1}^0f_{i+1})+A_i^0F_i$$

$$A_i^0n_i=A_i^{i+1}[A_{i+1}^0n_{i+1}+(A_{i+1}^0p_i^*)\times(A_{i+1}^0f_{i+1})] + (A_i^0p_i^*+A_i^0\hat{s}_i)\times(A_i^0F_i)+A_i^0N_i$$

$$\tau_i=\begin{cases} (A_i^0n_i)'(A_i^{i-1}z_0)+b_i\dot{q}_i, \text{ if link } i \text{ is rotational,} \\ (A_i^0f_i)'(A_i^{i-1}z_0)+b_i\dot{q}_i, \text{ if link } i \text{ is translational.} \end{cases}$$

[a] The subscript i stands for the ith link of the manipulator. The equations above the dashed line are calculated repeatedly for $i = 1$ to n (the number of links) and those below the line for $i = n$ to 1 in the reverse direction. A_i^{i+1} and J_i are 3 × 3 matrices; $\dot{q}_i$, $\ddot{q}_i$, m_i, b_i, and τ_i are scalars. The remaining variables are all 3 × 1 vectors.

Fig.1 Newton-Euler equations for robot-arm dynamic control.

2.2. Robot Dynamics Simulation

Robot dynamics simulation using Walker and Orin's algorithm also contains a heavy computational load composed of a lot of small-matrix operations, solution of linear equations and numerical integration. The algorithm includes a lot of 3 by 3 matrix operations and 3 by 1 vectors operations shown in Fig. 2. In Fig. 2, equations (1) to (12) calculate the bias vector, which is the summation of gravity, Coriolis, centrifugal and external forces. These equations are Newton-Euler equations. Let a joint (angular) displacement q and a (angular) velocity $\dot{q}$ be a current value and set a (angular) acceleration $\ddot{q}$ equal to zero. Then equations (1) to (8) are calculated repeatedly for i = 1 to the number of links of the manipulator n as indicated by "do i = 1 to n" and "end do" in the figure and equations (9) to (12) for i = n to 1 in the reverse direction. In equations (13) to (28), the inertia matrix H is computed. By solving the linear equations (29) using the bias vector and the inertia matrix H and the torques and the forces applied to the joints, the (angular) acceleration of each joint $\ddot{q}$ is computed. Then $\dot{q}$ and q at the next time instance are calculated by numerical integration.

do i=1 to n;

$$A_i^0\omega_i=\begin{cases}A_i^{i-1}(A_{i-1}^0\omega_{i-1}+z_0\dot{q}_i) & \text{(link i : Rotational) (1)}\\ A_i^{i-1}(A_{i-1}^0\omega_{i-1}) & \text{(link i : Translational) (2)}\end{cases}$$

$$A_i^0\dot{\omega}_i=\begin{cases}A_i^{i-1}(A_{i-1}^0\dot{\omega}_{i-1}+A_{i-1}^0\omega_{i-1}\times(z_0\dot{q}_i)) & \text{(R) (3)}\\ A_i^{i-1}(A_{i-1}^0\dot{\omega}_{i-1}) & \text{(T) (4)}\end{cases}$$

$$A_i^0\dot{v}_i=\begin{cases}(A_i^0\dot{\omega}_i)\times(A_i^0p_i)+(A_i^0\omega_i)\times((A_i^0\omega_i)\times(A_i^0p_i))\\ \qquad +A_i^{i-1}(A_{i-1}^0\dot{v}_{i-1}) & \text{(R) (5)}\\ A_i^{i-1}(A_{i-1}^0\dot{v}_{i-1})+(A_i^0\dot{\omega}_i)\times(A_i^0p_i)+2(A_i^0\omega_i)\times(A_i^{i-1}z_0\dot{q}_i)\\ \qquad +(A_i^0\omega_i)\times((A_i^0\omega_i)\times(A_i^0p_i)) & \text{(T) (6)}\end{cases}$$

$$A_i^0F_i=m_i\{(A_i^0\omega_i)\times((A_i^0\omega_i)\times(A_i^0s_i))+(A_i^0\dot{\omega}_i)\times(A_i^0s_i)+A_i^0\dot{v}_i\} \quad (7)$$

$$A_i^0N_i=(A_i^0J_iA_0^i)(A_i^0\dot{\omega}_i)+(A_i^0\omega_i)\times((A_i^0J_iA_0^i)(A_i^0\omega_i)) \quad (8)$$

end do;

do i=n to 1;

$$A_i^0f_i=A_i^{i+1}(A_{i+1}^0f_{i+1})+A_i^0F_i \quad (9)$$

$$A_i^0n_i=A_i^{i+1}(A_{i+1}^0n_{i+1}+(A_{i+1}^0p_i)\times(A_{i+1}^0f_{i+1}))+(A_i^0p_i+A_i^0s_i)\times(A_i^0F_i)+A_i^0N_i \quad (10)$$

$$bias_i=\begin{cases}(A_i^0n_i)^T(A_i^{i-1}z_0)+b_i\dot{q}_i & \text{(R) (11)}\\ (A_i^0f_i)^T(A_i^{i-1}z_0)+b_i\dot{q}_i & \text{(T) (12)}\end{cases}$$

end do;

$$M_n=m_n \quad (13)$$

$$A_{n-1}^0c_n=A_{n-1}^n(A_n^0s_n+A_n^0p_n) \quad (14)$$

$$A_{n-1}^0E_nA_0^{n-1}=A_{n-1}^n(A_n^0E_nA_0^n)A_n^{n-1} \quad (15)$$

do j=n-1 to 1;

$$M_j=M_{j+1}+m_j \quad (16)$$

$$A_{j-1}^0c_j=A_{j-1}^j(A_j^0c_j)=A_{j-1}^j(\frac{1}{M_j}\{m_j(A_j^0s_j+A_j^0p_j)+M_{j+1}(A_j^0c_{j+1}+A_j^0p_j)\}) \quad (17)$$

$$\begin{aligned}A_{j-1}^0E_jA_0^{j-1}=A_{j-1}^j(&A_j^0E_{j+1}A_0^j+A_j^0J_jA_0^j\\ &+M_{j+1}\{(A_j^0c_{j+1}+A_j^0p_j-A_j^0c_j)\cdot(A_j^0c_{j+1}+A_j^0p_j-A_j^0c_j)I\\ &-(A_j^0c_{j+1}+A_j^0p_j-A_j^0c_j)(A_j^0c_{j+1}+A_j^0p_j-A_j^0c_j)^T\}\\ &+m_j\{(A_j^0s_j+A_j^0p_j-A_j^0c_j)\cdot(A_j^0s_j+A_j^0p_j-A_j^0c_j)I\\ &-(A_j^0s_j+A_j^0p_j-A_j^0c_j)(A_j^0s_j+A_j^0p_j-A_j^0c_j)^T\})A_j^{j-1}\end{aligned} \quad (18)$$

$$A_{j-1}^0F_j=\begin{cases}z_0\times M_jA_{j-1}^0c_j & \text{(R) (19)}\\ (0\ 0\ M_j)^T & \text{(T) (20)}\end{cases}$$

$$A_{j-1}^0N_j=\begin{cases}(A_{j-1}^0E_jA_0^{j-1})z_0 & \text{(R) (21)}\\ 0 & \text{(T) (22)}\end{cases}$$

$$A_{j-1}^0f_j=A_{j-1}^0F_j \quad (23)$$

$$A_{j-1}^0n_j=A_{j-1}^0N_j+(A_{j-1}^0c_j)\times(A_{j-1}^0F_j) \quad (24)$$

do i=j-1 to 1;

$$A_{i-1}^0f_i=A_{i-1}^i(A_i^0f_{i+1}) \quad (25)$$

$$A_{i-1}^0n_i=A_{i-1}^i(A_i^0n_{i+1}+(A_i^0p_i)\times(A_i^0f_{i+1})) \quad (26)$$

end do;

end do;

do j=n to 1;

do i=j to 1;

$$H_{ij}=\begin{cases}(A_{i-1}^0n_i)_z & \text{(R) (27)}\\ (A_{i-1}^0f_i)_z & \text{(T) (28)}\end{cases}$$

end do;

end do;

$$H\ddot{q}=\tau-bias \quad (29)$$

$$\ddot{q}\rightarrow\dot{q},\ q$$

where . n : Number of links; b_i, M_i, m_i : scalar

A_i^j, J_i, E_i, I : 3×3 matrix; H : n×n matrix

$\omega_i, \dot{\omega}_i, v_i, \dot{v}_i, p_i, s_i, F_i, N_i, f_i, n_i, c_j$: 3×1 vector

$\tau, bias, \ddot{q}, \dot{q}, q$: n×1 vector; $z_0=(0\ 0\ 1)^T$

Fig.2 Walker & Orin's algorithm for dynamics simulation.

3. PARALLEL PROCESSING SCHEME

This section presents the parallel processing scheme which allows us to execute control and simulation computation for an arbitrarily-shaped robot-arm in the minimum time on a multiprocessor system with a arbitrary number of processors. Key points for parallel processing on a general-purpose multiprocessor system are how to decompose the computation into several small parts of computation called tasks and how to allocate or schedule the tasks to processors.

3.1 Generation of tasks

The first step of the parallel processing scheme is to decompose the control computation into tasks, which are small units of computation to be allocated to a processor, so that the parallelism can be effectively extracted from the Newton-Euler equations and the Walker & Orin's algorithm.

In the decomposition, we should determine the optimal task size from the following several possible task sizes, or grains.

The first one is an equation level grain in which each equation is allocated to a processor as a task. The second one is a matrix or vector operation level grain in which one or several matrix/vector operations are treated as a task. The last one is a scalar operation level grain in which one or several arithmetic operations correspond to a task.

The determination of the task size has a strong influence on the performance of parallel processing [16][27]. At a glance, the smallest task size seems to give us the shortest parallel processing time since it can extract the maximum parallelism. In practice, however, it does not always give us the shortest processing time because it suffers from large processing overheads such as data transfer overhead and task synchronization overhead. On the contrary, if a large task size is chosen, only limited parallelism can be extracted, but the overheads are relatively small. Therefore, in order to get the minimum processing time, the task size must be decided taking into account the following factors;

a) processing capability of each processor element,
b) data transfer capability among processor elements,
c) overhead caused by synchronization,
d) capability of a scheduling algorithm.

Here, the equation level task grain is chosen by taking into consideration the above factors of an multiprocessor system shown in the next section although the parallel processing scheme can equally be applied to the other task grains.

When we decompose Newton-Euler equations for the Stanford manipulator with six D.O.F. into tasks by using equation level task grain, eighty-eight tasks are generated as Luh and Lin[1].

Also, in the task generation for the simulation, the equation level task grain is basically chosen. However, in a part for the solution of the linear equation using the Gauss-Jordan method with the pivot selection, dividing operations by the pivots and up-dating operations for each row are treated as a task to generate relatively coarse grain tasks. For a part of the numerical integration using Euler, Trapezoidal or 4-th Adams-Bashforth method, integration calculation for each variable is treated as a task.

Furthermore, in the task generation of the simulation, program restructuring techniques like the array (scalar) expansion performed by parallelizing compilers[31] is used to extract potential parallelism of the simulation program[10]. The array expansion technique allows us to parallelize a program part which must be executed serially because of multiple definitions of a vector variable, namely output data dependences[31], by expanding the vector variable into a two-dimensional array variable[10].

In the simulation, a user may frequently change a configuration or various parameters of the target robot-arm. In this case, a simulation program must be re-generated for each change. However, since the change of the program is a time- consuming job for users, a special purpose compiler including a preprocessor should automatically generate the simulation program and decompose the program into tasks by using the above mentioned method.

3.2 Generation of Task Graph

There exist precedence constraints among the generated tasks caused by data dependences[18] among them. In other words, if a vector variable defined by task i is referred by another task j, task j must wait completion of task i. The precedence constraints, or the execution order constraints, among the tasks can be represented by a directed acyclic graph called a task graph as shown in Fig. 3. Fig. 3 represents the task graph with 88 tasks for the Newton-Euler equations for Stanford manipulator. In the task graph, the nodes represent the tasks and the arcs among nodes show the precedence constraints. For example, arcs among nodes 1, 4 and 6 mean that task 6 must wait to begin execution until tasks 1 and 4 are completed. A processing time of each task on a processor is attached beside the node. However, task processing time is not always a deterministic value since a processing time for a floating point operation may vary with values of operands, or data. In light of this fact, an average task processing time is used for the scheduling described later.

Once the task graph is generated by the data-flow and control flow analysis among tasks, the lower bound of parallel processing time for the tasks can be obtained as a critical path length t_{cr} of the task graph[11][19]. Thus the most efficient multiprocessor system is the one which can attain t_{cr} with the minimum number of processors because we can't obtain a shorter processing time than t_{cr} even if more processors were used.

Fig. 4 represents a task graph with 197 tasks for Walker & Orin's algorithm for Stanford Manipulator though task processing times are omitted.

3.3 Task scheduling

In order to process the set of tasks on a multiprocessor system efficiently, an assignment of the tasks onto the parallel processors and an execution order of the tasks assigned to the same processor must be determined optimally. The problem which determines the optimal task assignment and the optimal execution order can be treated as the traditional multiprocessor scheduling problem of which the objective function is the minimization of the execution time or the schedule length[11][17][19]. The

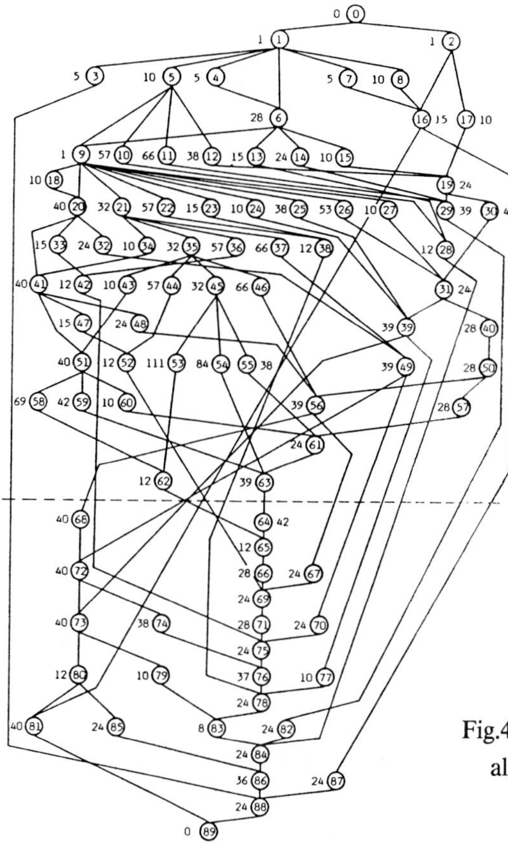

Fig. 3. A task graph for Newton-Euler equations.

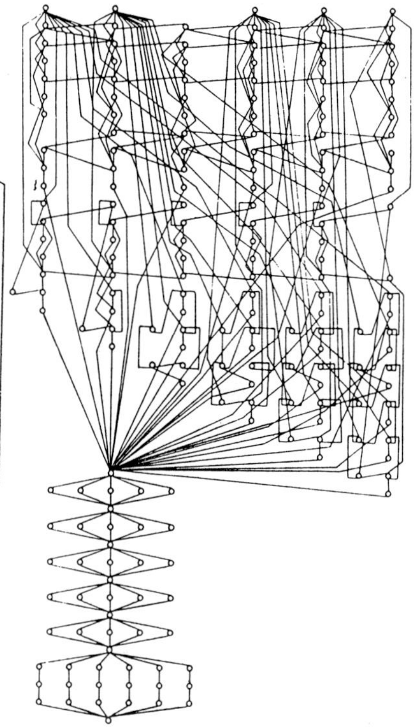

Fig.4. A task graph for Walker & Orin's algorithm.

scheduling problem, however, has been known as one of the most intractable optimization problems, for which no practical optimization algorithms had been developed for near 20 years. More precisely speaking, the problem belongs to the class of "strong" NP-hard problems[21]. In order to decrease the difficulty, Luh and Lin intuitively determined the assignment of the tasks to processors by fixing the number of processors at six, which is equal to the number of joints of Stanford manipulator. In other words, they assigned tasks related to each joint onto a processor[1]. After this simplification, they tried to optimize only the execution order of the tasks assigned to each processor by taking into consideration the precedence constraints among tasks assigned to its neighboring processors. Finally, they obtained a schedule which reduces the processing

Table 1 Comparison of schedules by Luh's method and Kasahara's method

No. of Processors	Processing time [ms]	
	Luh & Lin	Kasahara & Narita
1	24.80	24.83
2		12.42
3		8.43
4		6.59
5		5.86
6	9.67	5.73
7		5.69

time for the eighty-eight tasks from 24.8[ms] for a single processor to 9.67[ms] for six processors as shown in Table 1.

On the other hand, Kasahara & Narita successfully solved the original scheduling problem by developing two practical algorithms named CP/MISF (Critical Path / Most Immediate Successors First) and DF/IHS (Depth First / Implicit Heuristic Search)[11]. The former is one of list scheduling algorithms [19][22], or a kind of heuristic scheduling algorithms, and quickly gives us accurate approximate solutions even on a microcomputer.

The algorithm of CP/MISF method is of the following.

Step1: Determine the "level" for each task, the level l_i of task i being defined as the longest path from the exit node to node N_i by the $O(n^2)$ algorithm using a feature of directed acyclic graph[24].

Step2: Construct the priority list in the descending order of l_i and the number of immediately successive tasks.

Step3: Execute list scheduling [19][22] on the basis of the priority list.

The worst-case performance of CP/MISF, i.e., the error of the worst-case solution t obtained by CP/MISF from the optimal solution t_{opt} is given by $t(2\text{-}1/m)t_{opt}$. Time complexity of CP/MISF is $O(n^2 + mn)$, where O means the order function, n is the number of tasks and m is the number of processors. Since CP/MISF is a heuristic algorithm, it is difficult to know if the solution is optimal.

DF/IHS is a powerful optimization algorithm which combines CP/MISF and the branch & bound method [23][25] in a special manner and allows us to obtain optimal solutions or approximate solution with guaranteed accuracy. The solutions given by DF/IHS are always more accurate than those by CP/MISF. The proper use of the both algorithms depends on the required accuracy of the solution, the processing capability of the computer to be used and the number of tasks to be scheduled. More details about these algorithms and their extended versions considering data transfer overhead has been appeared in references[11] and [28].

By using DF/IHS algorithm, the original problem to schedule the eighty-eight tasks for Stanford manipulator to an arbitrary number of processors can be solved. Table 1 lists the scheduled results by DF/IHS, where the number of processor is varied from one to seven. The schedules by DF/IHS are optimal or near optimal for every number of processors. Also, the schedule length, or execution time, for three processors is shorter than Luh's schedule for 6 processors. Furthermore, if six processors are used, the processing time is reduced by 40 percent. The Gantt chart in Fig. 5 shows an optimal schedule for three processor. In this figure, P1, P2 and P3 means processor number, horizontal axis of each row shows elapsed time and figures inside a rectangle represent task number. From the figure, we realize that an optimal schedule sometimes disagrees with or even contradicts human intuition.

Also, it has been already shown that the scheduling algorithm can effectively be applied for another manipulator and another control scheme like the kinetic control scheme.

In terms of scheduling of the Newton-Euler equations, Bahren, Toomarian and Protopopescu has showed that an application of a neural network to the multiprocessor scheduling gives us almost the same length of a schedule for four processors as DF/IHS algorithm[20].

DF/IHS and CP/MISF algorithms do not consider data transfer overhead among processors because relatively coarse task grain has been chosen taking into consideration capabilities of multiprocessor systems in hand. However, scheduling algorithms should take into consideration data transfer overhead when relatively fine task grain is chosen or a multiprocessor system having low data transfer capability is used. However, the scheduling problem which minimizes the execution time taking into account data

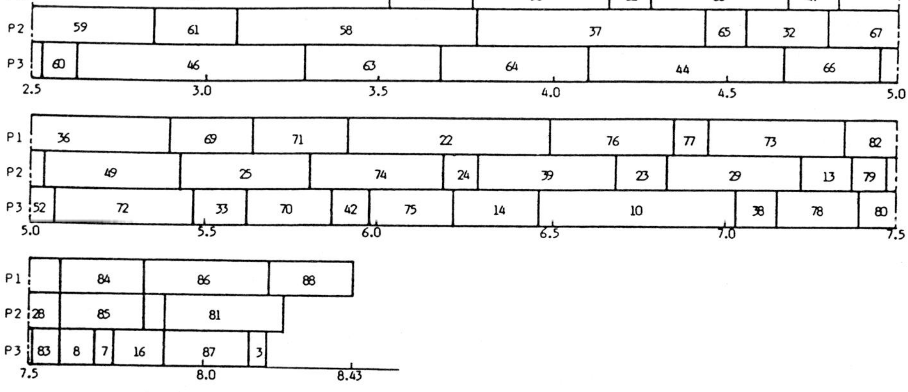

Fig.5. An optimal schedule for Newton-Euler equations for three processors given by DF/IHS algorithm.

transfer is an extremely intractable problem. For example, no practical optimization algorithm has not been proposed. Therefore, a heuristic algorithm is most promising for this problem. As a practical heuristic algorithm, CP/DT/MISF (CP/ Data Transfer/ MISF) algorithm, which is an extended algorithm of CP/MISF, has been proposed for the problem considering data transfer[32].

3.4 Generation of Parallel Machine Code

Next, a parallel machine code is generated by using a scheduled result. In this section, a code generation scheme for a typical shared memory multiprocessor system, in which processor elements (PEs) having local memories are connected to a common memory by a bus as shown in section 4, is described.

The machine code for each PE is generated by putting instructions for tasks assigned to the PE in the order specified by a scheduled result and inserting synchronization instructions and data transfer instructions before or after tasks which need data transfers to other processors.

The instruction sequence for a task is optimized so that the registers in a processor can be used as much as possible. The synchronization among tasks can be realized by the version number method, which does not need a hardware support for synchronization, because the synchronization problem in the present case becomes "one-writer / multiple-reader problem". In the version number method, a writer task, which writes the output data of the task (the shared data) onto the common memory (CM) in order to send the data to other tasks allocated to other processors, updates the version number on the CM after writing the shared data. The reader tasks then confirm the update of the version numbers and read the data.

A version number is assigned to each shared data on the CM and the version number is a function of a number of the iterations which has already been processed.

Every PE has the same version number during one iteration and the version number is increased by each PE independently at the end of the iteration. In code generation, instructions to update the version number on CM are inserted after machine instructions for each writer task and instructions to check the update of the version number on CM are inserted before instuructions for each reader task.

The update and check of version numbers causes overhead. The overhead, however, can be minimized by removing redundant synchronization instructions by using information obtained from the scheduled result.

For example, let tasks A, B and C be allocated to processor 1 and tasks D and E to processors 2 and 3 respectively as shown in Fig.6. Then, task B does not need to check the version number set by task A. This stems from the fact that both tasks are allocated to the same processor and task C has no need to check the version number set by task D since the termination of task D has been confirmed by task B. Furthermore, the frequency of flag check by a reader task is reduced if it is estimated from the schedule that a reader task must wait for the set of the version number for a long time because the frequent flag checks cause a large overhead.

The task generation, the scheduling and the machine code generation are automatically performed by the dedicated compiler.

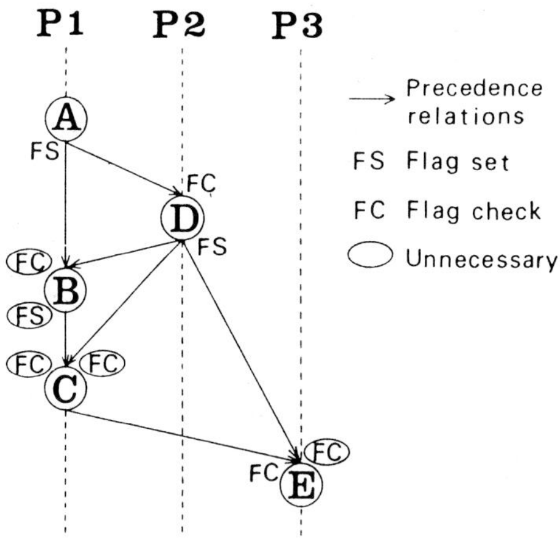

Fig. 6. Minimization of synchronization overhead.

4. PARALLEL PROCESSING ON AN ACTUAL MULTI-MICROPROESSOR SYSTEM

4.1 An Experimental Multiprocessor System

Fig. 7 shows a multi-microprocessor system which was developed in 1984 to evaluate the practicality of the software oriented parallel processing scheme using the static

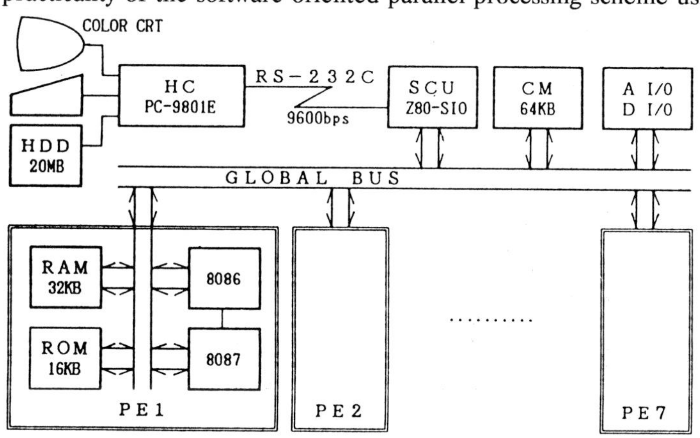

Fig. 7. A multimicroprocessor system used for evaluation of the software oriented parallel processing scheme.

scheduling at compile-time[4][10] mentioned in the previous section. This system consists of seven Processor Elements (PEs), a Common Memory (CM), a single bus, and a Serial Communication Unit (SCU) to control the data transfer between a Host Computer and PEs. Each PE is composed of a 16-bit microprocessor intel 8086, its co-processor for floating-point operations 8087 and a local memory (16KB ROM, 32KB RAM). The data transfers among the PEs is made via the common memory. The bus access control is made by a distributed bus arbiter on each PE or SCU. Each arbiter controls the bus accesses based on a priority assigned a priori to each PE and SCU. The HC is a 16-bit personal computer and connected to the multiprocessor unit by a serial communication link (RS-232C).

4.2 Parallel Processing of Dynamic Control Computation

This section introduces performance of parallel processing of Newton-Euler equations for Stanford manipulator.

Curve A in Fig. 8 shows the optimal schedule length for one to seven processors obtained by DF/IHS, namely a theoretical processing time, and Curve B represents the processing time measured on the actual multiprocessor system. The theoretical processing time means the minimum processing time when a given number of processors are used. More precisely speaking, it is an ideal (shortest) processing time for each number of processors without overheads caused by bus contentions and a synchronization between iteration steps. In five processors, Curve A reaches the critical path length. This fact shows that the use of more than five processors is unnecessary when the equation level task grain is adopted on the experimental multiprocessor system. Also, Curve B is very close to the ideal curve. This fact shows that efficient parallel processing with low overhead was realized on the actual multiprocessor system.

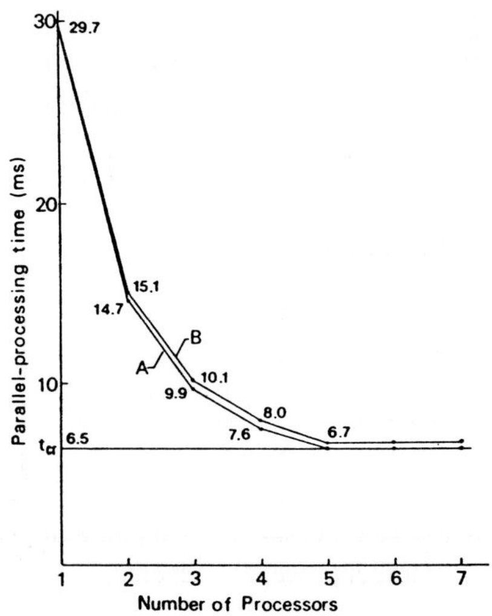

Fig. 8. Parallel processing time of Newton-Euler equations vs. number of processors.

In addition, the measured processing time is reduced from 29.7[ms] for single processor to 15.1[ms] for two processors, 10.1[ms] for three processors and 6.7[ms] for five processors which is almost t_{cr}. This processing time of 6.7[ms] easily clears the longest sampling period of 1/60[s] for real time operation stated by Luh[1].

These results show the parallel processing scheme using the static scheduling allows us to obtain the minimal parallel processing time for any number of processors.

4.3 Performance of Parallel Processing of Dynamics Simulation

This section describes performance of parallel processing of dynamics simulation using Walker & Orin's algorithm for Stanford manipulator.

Curve A in Fig. 9 shows the optimal schedule obtained by DF/IHS and Curve B shows the processing time measured on the actual multiprocessor system.

Curve A reaches the critical path length, in six processors.

Curve B is also very close to the ideal curve.

Furthermore, the measured processing time shown by Curve B is reduced from 84.97[ms] for single processor to 44.40[ms] for two processors, 30.24[ms] for three processors and 17.93[ms], which is near t_{cr}, for six processors. This processing time of 17.93[ms] almost clears the longest sampling period of 1/60[s] for real time operation although the used multiprocessor system has very small processing capability. In addition, the processor utilization factor is very high; 96% for two processors, 94% for

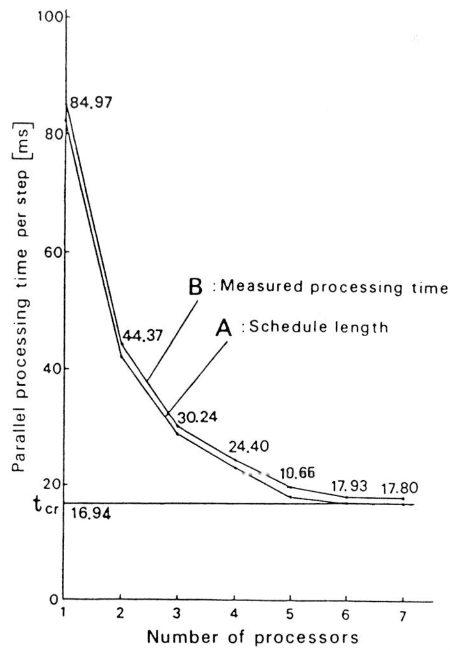

Fig. 9. Parallel processing time of dynamics simulation vs. number of processors.

three processors and 79% even for six processors by which the critical path length is almost attained.

These results show the parallel processing scheme mentioned in the previous section is practical and effective.

5. CONCLUSIONS

This chapter has introduced a parallel processing scheme using the static multiprocessor scheduling algorithms for the robot-arm dynamic control computation using Newton-Euler equations and robot-arm dynamics simulation using Walker-Orin's algorithm. In this scheme, if an arbitrary robot configuration is input, generation of control or simulation programs for robot dynamics, task generation, data flow analysis, generation of task graph, scheduling, machine code generation, parallel execution can be performed automatically by the dedicated compiler. Also, practicality and usefulness of the scheme were shown on an actual multi-microprocessor system.

The results show that a cost-effective multi-microprocessor robot controller and simulator can be put into practical use soon. The realtime multi-microprocessor robot simulators will allow us to efficiently design a more complex robot configuration and to evaluate performance of a new controller by directly connecting the simulator to the controller.

Furthermore, the proposed parallel processing scheme can be easily applied for processing of parallel algorithms[29], which has been actively researched to fully use high processing capability of a multiprocessor system and to minimize data transfer and synchronization overhead among processors.

Finally, it is expected that a multi-microprocessor robot controller will be more important to realize a compact and powerful controller to control an advanced autonomous robot system having various sensors including vision and intellectual functions using a neural network.

REFERENCES

[1] J.Y.S.Luh and C.S.Lin, "Scheduling of parallel computer for a computer controlled mechanical manipulator," IEEE Trans. SMC, Vol 12, No 2,pp.214-234,Mar.1983.

[2] H.Kasahara and S.Narita, "Parallel processing scheme for robot control computation using multiprocessor scheduling algorithms," J. of the Robotics Soc. of Japan, vol.2, No.5, pp.387-400, Oct. 1984 (in Japanese).

[3] D.E.Orin, "Pipelined approach to inverse plant plus Jacobian control of robot manipulators," in Proc. IEEE Int. Conf. Robotics, pp.169-175, 1984.

[4] H.Kasahara and S.Narita, "Parallel processing of robot arm control computation on a multimicroprocessor system," IEEE J. of Robotics and Automation, Vol. RA-1, No.2, pp.104-113, June 1985.

[5] R.H.Lathrop, "Parallelism in manipulator dynamics," Int. J. of Robotics Research, vol.3, pp.80-102, 1985.

[6] D.E.Orin, H.H.Chao, K.W.Olson and W.W.Schrader, "Pipeline / parallel algorithms for the Jacobian and inverse dynamics computations," in Proc. 1985 IEEE Int. Conf. on Robotics and Automation, pp.785-789, 1985.

[7] F.Y.Liao and M.Y.Chern, "Robot manipulator dynamic computation on a VLSI array processors," in Proc. 1st IEEE Int. Conf. on Supercomputing Systems, pp.116-121, Dec. 1985.

[8] R.Nigan and C.S.G.Lee, "A Multiprocessor-Based Controller for the Control of Mechanical Manipulators," IEEE J. Robotics and Automation, Vol. RA-1, No.4, pp.173-182, Dec. 1985.

[9] M.K.Vukobratovic and L.M.Kircanski, "An approach to parallel processing of dynamic robot models," in Proc. 10th World Congress, Pergamon Press: Oxford, July 1987.

[10] H.Kasahara, H.Fujii and M.Iwata, "Parallel processing of robot motion simulation," in Proc. 10th World Congress, Pergamon Press: Oxford, Jul. 1987.

[11] H.Kasahara and S.Narita,"Practical multiprocessor scheduling algorithms for efficient parallel processing," IEEE Trans. Comput. Vol. C-33, No.11, pp.1023-1029, Nov. 1984.

[12] J.Y.S.Luh, M.W.Walker and R.P.C.Paul, "On-line computational scheme for mechanical manipulators," ASME J. Dynamic Syst., Meas., Contr., Vol. 102, pp.69-79, June 1980.

[13] M.W.Walker and D.E.Orin, "Efficient dynamic computer simulation of robotic mechanism," in Proc. Jt. Automatic Control Conf., 1981.

[14] J.M.Hollerbach, "A recursive formulation of Lagrangian manipulator dynamics," in Proc. Jt. Automatic Control Conference, 1980.

[15] S.Megahed and M.Renaud, "Minimization of the computation time necessary for the dynamic control of robot manipulators," in Proc. 12th Int. Symposium on Industrial Robots, pp.469-478, 1982.

[16] D.D.Gajski and J.K.Peir, "Essential issues in multiprocessor systems," Computer, Vol.18, No.6, pp.9-27, June 1985.

[17] H.Kasahara and S.Narita, "An approach to supercomputing using multiprocessor scheduling algorithms," in Proc. IEEE 1st Int. Conf. on Supercomputing Systems, pp.139-148, Dec. 1985.

[18] D.A.Padua, D.J.Kuch and D.H.Lawrie, "High-speed multiprocessors and compilation techniques," IEEE Trans. Comput., Vol. C-29, pp.763-776, Sep. 1980.

[19] E.G.Coffman, Computer and Job-shop Scheduling Theory. New York: Wiley, 1976.

[20] J.Bahren, N.Toomarian and V.Protopopescu, "Optimization of the computational load of a hypercube supercomputer onboard a mobile robot," Applied Optics, Vol. 26, No. 23, pp 5007-5014, Dec. 1987.

[21] M.R.Garey and D.S.Johnson, Computers and Intractability: A Guide to the Theory of NP-Completeness. San Francisco: Freeman, 1979.

[22] T.L.Adam, K.M.Chandy and J.R.Dickson, "A comparison of the list schedules for parallel processing systems," Commun. Ass. Comput. Mach., Vol.17, pp.685-690, Dec. 1974.

[23] T.Ibaragi, Combinatorial Optimization. Tokyo: Sangyo Tosyo, 1979.

[24] E.L.Lawler, Combinatorial Optimization: Networks and Matroids. New York: Holt, Rinhart and Winston, 1976.

[25] W.H.Kohler, "Characterization and theoretical comparison of branch-and-bound algorithms for permutation problems," J. Ass. Comput. Mach., Vol.21, pp.140-156, Jan. 1974.

[26] E.B.Fernandez and B.Bussel, "Bounds or the number of processors and time for multiprocessor optimal schedules," IEEE Trans. Comput., Vol.22, pp.745-751, Aug. 1973.

[27] B.Kruatrachue and T.Lewis, "Grain size determination for parallel processing," IEEE Software, Vol.5 , No 1, pp23-32, Jan.1988.

[28] H.Kasahara and S.Narita, "Load distribution among real- time control computers connected via communication media," in Proc. IFAC 9th World Congress, Pergamon Press, July 1984.

[29] A.Fijany and A.K.Bejczy, " A Class of Parallel Algorithms for Computation of the Manipulator Inertia Matrix," IEEE Trans. Robotics & Auto., Vol. RA-5, No.5, pp.600-615, Oct. 1989.

[30] H.Kasahara, H.Honda and S.Narita,"Parallel Processing of Real-time Dynamic Systems Simulation on OSCAR (Optimally SCheduled Advanced multiprocessoR), in Proc. NASA DOD NSF Conf. on Aerospace Control Computation, Oxnard:California, Aug. 1989.

[31] D.A.Padua and M.Wolfe,"Advanced Compiler Optimization for Supercomputers," C.ACM, Vol.29, No.12, pp.1184-1201, Dec. 1986.

[32] H.Kasahara et.al., "Parallel Processing of Near Fine Grain Tasks Using Static Scheduling on OSCAR," in Proc. IEEE ACM Supercomputing'90, Nov. 1990.

CHAPTER 4

ASPARC: AN ALGORITHMICALLY SPECIALIZED PARALLEL ARCHITECTURE FOR ROBOTICS COMPUTATIONS

Amir Fijany and Antal K. Bejczy

Jet Propulsion Laboratory
California Institute of Technology
Pasadena, California

Abstract

ASPARC, an algorithmically specialized parallel architecture is presented in this chapter for fast computation of a class of kinematic and dynamic problems required for control and simulation of robot manipulators. This class of problems include forward kinematics, Jacobian and its time derivative, inverse and forward dynamics, and operational space mass matrix and its inverse. It is shown that for the problems of this class there are both time and processors optimal parallel algorithms with $O(\log n)$ computational complexity which can be implemented on a unique parallel architecture. It is also shown that a significant speedup in computation of these problems can only be achieved by exploiting parallelism at several computational levels. Therefore, the design of ASPARC is optimizied for a multilevel parallel computation of this class of problems. Parallel implementation of Resolved Acceleration Control (RAC) scheme on ASPARC is analyzed and it is shown that speedups of about 16.3 and 21.9 for 6 and 8 degree-of-freedom robot arms can be achieved. A fast technique for both serial and parallel computation of time derivative of Jacobian, which is needed for implementation of RAC, is presented.

I. Introduction

In recent years, several model-based control schemes have been proposed to linearize and decouple the equations of motion of robot manipulators [4,18,25,34,43]. The task space control schemes, by taking into account the kinematic and dynamic properties of the manipulator, are particularly expected to achieve significant improvement over the conventional independent joint control schemes. However, this improvement is achieved at the expense of a significant increase in the computational requirements for control.

As a result, the real-time implementation of these control schemes requires hardware architectures with significant computation power. On one hand, the computation should be performed at a high rate since the stability of these control schemes requires a sampling rate at least 10 times the structural resonant frequency of the manipulator. On the other hand, at each sampling interval a set of computation intensive kinematic and dynamic problems need to be solved. The Resolved-Acceleration Control (RAC) [34] represents an illustrative example of computational requirements of the task space control schemes. The implementation of RAC requires the computation of forward kinematics, Jacobian, its time derivative and its inverse, and inverse dynamics at each sampling interval. The Operational Space Control [25] is even more computationally demanding since in addition to inverse dynamics, it requires the computation of operational space mass matrix and its inverse which are more computation intensive than the inverse dynamics.

The dynamic simulation of manipulator motion (or, the forward dynamics problem) is even more computation intensive than the task space dynamic control. The motivation for fast computation of forward dynamics solution stems from applications which require extensive off-line simulation as well as applications which require real-time simulation capability. In particular, for many anticipated space telemanipulation applications a faster-than-real-time dynamic simulation capability will be essential for achieving dexterous telemanipulation. In fact, in the presence of unavoidable delay in information transfer, this would enable the predictive mode control, i.e., the human operator would be able to preview several scenarios before run-time.

The quest for achieving dexterous manipulation has led to an increasing application of redundant arms as well as multiple cooperating arms. This increase in the system complexity also results in an increase in computational requirements for dynamic control and simulation. However, while the control and simulation of a redundant arm mostly represents an increase in the problem size, the simulation of multiple cooperating arms results in a significant algorithmic change. This is due to the fact that, during the same manipulation, the topology of the system can change from multiple serial open-loop chains, whose simulation requires the computation of classical forward dynamics problem [2,6,44,49], to a single closed-loop chain, whose simulation is completely different and requires the computation of operational space mass matrix and its inverse for each arms as well as the coupling among the arms [39,45,46].

In this paper, we consider a class of kinematic and dynamic computation problems which are required for dynamic control and simulation of open- and closed-chain arms. This class includes forward kinematics, Jacobian and its time derivative, inverse and forward dynamics, and operational space mass matrix and its inverse. We exclude the problem of Jacobian inversion since, unlike other problems which are generic and deterministic, its solution is highly task and arm dependent.

During the past two decades much research effort has been made to develop efficient serial algorithms for the class of problems considered above. This has resulted in a significant improvement in the efficiency of algorithms and the discovery of asymptotically optimal algorithms for all problems of this class. In fact, it seems unlikely that any but minor improvement in the efficiency of the serial algorithms can be achieved. However, despite their relative efficiency, the serial algorithms can not provide the computational efficiency required for real-time control and, in particular, for the faster-than-real-time simulation.

The maturity of serial algorithms and the advent of VLSI technology clearly indicates that the exploitation of parallelism in the computation is the key enabling factor for achieving the required computational efficiency. Consequently, many parallel algorithms and architectures have been proposed for this class of problems, and a few reports discussed the implementation of the proposed algorithms and architecture to practically validate their efficiency. Usually a specific problem, e.g., parallel computation of Jacobian [16,41,42] or inverse dynamics [3,11,24,30,32,36,51], are considered while, as discussed before, both dynamic control and simulation are multiphase problems in the sense that their implementation requires efficient computation of a set of problems. In fact, in addition to cost-effectiveness issue, separate design of most efficient parallel algorithm and architecture for each specific problem may not result in an overall efficiency since the sequencing of and communication among different architectures can become the major bottleneck in the computation. (See for example [12,31] for parallel computation of forward dynamics).

The question then arises as whether there is a unique optimal parallel architecture for control and simulation problems. In other words, the question is whether it is possible to design a unique architecture on which the class of problems considered here can all be efficiently implemented. In addition to its cost-effectiveness, such an architecture will provide a practical solution for high performance parallel computation of both control and simulation problems. It can also provide further computational efficiency by reducing the overhead due to sequencing and chaining of various computations, on one hand, and by exploiting the synergism (common computation) among them, on the other hand.

In order to achieve maximum computational efficiency, the architecture can be specifically optimized for this class of problems. In this sense, it can be an *application-specific* or an *algorithmically-specialized parallel architecture*. The design of such an optimal architecture for this class of problems raises two challenging issues. From an algorithmic point of view, the challenge is to design optimal parallel algorithms for this class of problems with similar properties. Note that these are the structural properties of the parallel algorithms, e.g., the algorithm's communication and synchronization structure, granularity, type of operations, etc., which are of interest since they indicate the architectural requirements for the efficient implementation of the algorithm [22,47]. From an architectural point of view, the challenge is then to design an architecture with managable complexity to exploit the algorithmic properties.

Algorithmically-specialized parallel architecture represents an approach to parallel computation that requires the analysis of algorithmic and architectural issues and their relationships. Generally, there are two common approaches to parallel computation of a given problem. In the first approach, a target parallel architecture is considered initially, the features of which define the parallel model of computation in advance. An algorithm is then developed to exploit the parallelism in the problem by taking into account the given parallel computation model. The drawback of this approach is that the architectural features can limit the algorithm's performance. The limitation can result from cardinality variation (i.e., the difference between the desired algorithm's processes and the architecture's processors) and the topological variation (i.e., the difference between the communication structure inherent to the algorithm and that of the architecture). More importantly for our case here, the architectures considered are usually capable of exploiting only one type of parallelism at one computational level which may not be the most significant exploitable parallelism for the problem.

When performance is the main criteria, a second approach is chosen in which an algorithm is first developed to efficiently exploit the parallelism in the problem and then an algorithmically specialized architecture is designed to fully support the algorithm. The challenge in this approach resides in both the algorithmic and architectural design. Given a problem, several parallel algorithms for its computation can be developed. However, the real efficiency of any algorithm depends on its suitability for architectural optimization. Hence, the main challenge is not to develop any algorithm but to develop the algorithms with structural properties amenable to architectural optimization.

In this chapter, we present the design of an **A**lgorithmically **S**pecialized **P**arallel **A**rchitecture for **R**obotic **C**omputations (designated as **ASPARC**) which provides the optimal performance for this class of problems. The main contribution of this chapter is to show that not only there are asymptotically optimal (in terms of both computation time and number of processors) parallel algorithms for the class of problems with similar structural properties but that these properties also are amenable to architectural optimization. We also show that, in order to achieve significant speedup in the different computations, parallelism needs to be exploited at several computational levels. Consequently, ASPARC is designed for a multilevel exploitation of parallelism in the class of problems.

It should be mentioned that the work of this chapter has been originated from our early works in [10,11] where an algorithmically-specialized architecture was developed for a subclass of these problem (i.e., forward kinematics, Jacobian, and inverse dynamics). The practical implementation of the architecture and the algorithms proved the validity of the approach and also provided insights into further architectural improvements which are reflected in the design of ASPARC. Also, recent progress in the design of optimal parallel algorithms for forward dynamics and operational space mass matrix and its inverse proved the efficiency of the architecture for these problems, thus making it optimal for the class of considered problems.

This chapter is organized as follows. In Sec. II, the parallelism in the class of problems is discussed. It is shown that there are both time and processors parallel algorithms for all the problems of this class with similar algorithmic properties and that they can be implemented on a unique architecture. The importance of multilevel exploitation of parallelism is also discussed. In Sec. III, various aspects of the design of ASPARC are discussed. In Sec. IV, implementation of some problems on the architecture is discussed and the resulting performance is evaluated. Some concluding remarks are made in Sec. V.

II. Parallelism in Kinematics and Dynamics Computations

A. Classification of Parallel Algorithms: Time and Processors Bounds in Computation

The analysis of time and processors bounds in parallel computation of a given problem is of fundamental theoretical importance. It can determine the maximum parallelism (i.e., time lower bound) in the problem and the maximum number of processors (i.e., processors bound) required for its exploitation. In addition to the theoretical importance, it can also provide useful insights for devising more practical and efficient parallel algorithms (in terms of both computation time and number of processors) for the problem. Parallel computation of forward dynamics, which is discussed below, represents such a case.

Let P denote the class of problems that can be solved sequentially in a time bounded by a polynomial of the input size, n. Also, let NC (for Nick's Class [27]) stand for the class of problems solvable in a polylogarithmic parallel time, i.e., in a time of $O(\log^k n)$, for some constant k, with a number of processors bounded by a polynomial of n. For $k = 1$, the time of $O(\log n) + O(1)$ represents the natural time lower bound in the computation. One open question regarding the complexity of parallel algorithms is whether $P = NC$, which is thought (though there is no formal proof) to be very unlikely [19] (It is clear that $NC \subseteq P$). The importance of this question resides in its relation to determination of the nature and amount of parallelism in computation. If a problem from P with the serial computational complexity of $O(n^s)$ belongs to NC then a speedup of $O(n^s/\log^k n)$ in its computation can be achieved. This indicates an *unbounded parallelism* and a speedup that grows significantly with the size of problem. For this reason the NC problems are considered as *well-parallelizable* problems [19].

Given an application domain, a related question is then whether all the problems from the application domain, representing a subset of P, belong to NC. The answer to this question, which determines the parallelism in the application domain, might be simpler than the general case since, most often, the problems from a given application domain exhibit similar algorithmic properties.

Interestingly, this is actually the case for the class of problems considered in this chapter since it is shown that they all belong to the class of NC. Furthermore, it is possible to devise parallel algorithms to achieve the time lower bound of $O(\log n) + O(1)$ for these problems. For example, forward kinematics and Jacobian [16,42], the time derivative of Jacobian (see Appendix), inverse dynamics [30,32], and the mass matrix [14] can be computed in $O(\log n) + O(1)$ steps. These algorithms are not only asymptotically optimal in terms of both computation time and number of processors but are also practical since their implementation requires a rather simple architecture. The parallel $O(\log n)$ algorithms for forward kinematics, Jacobian and its time derivative, and inverse dynamics can all be efficiently implemented on a parallel SIMD architecture with n processors interconnected through a Shuffle-Exchange (SE) augmented with Nearest Neighbor (NN) connection network (designated as SENN).

For a serial chain robot arm, the parallel algorithms for computation of forward kinematics, Jacobian, inverse dynamics, and mass matrix indeed reflect the common algorithmic properties of these problems which results from the recursive mechanical structure of the robot arm (i.e., a $(n+1)$-links arm is constructed by simply adding a link to a n-links arm). This recursive structure allows the representation of these problems as the solution of a set of first-order linear recurrences which can be computed in parallel in $O(\log n)$ steps [26].

Until recently very few results have been presented for parallel computation of forward dynamics for a serial open-chain arm and for parallel computation of operational space mass matrix and its inverse (and, hence, forward dynamics for closed-chain arm). The main obstacle in parallelization of the $O(n)$ algorithms for both open- and closed-chain arm forward dynamics is that the recurrences for computation of articulated-body inertia [6] and the operational mass matrix [39,45,46] are highly serial. In [12] it is shown that these recurrences can be represented as a set of first-order nonlinear recurrences of degree 2 which are well known to have a bounded parallelism [40]. That is, regardless of the number of processors, only a constant speedup in their parallel computation can be achieved. As a result, parallelization of these serial $O(n)$ algorithms result in yet $O(n)$ parallel algorithms which are faster than their serial counterparts by only a (small) constant factor.

Consequently, most of the proposals for parallel computation of forward dynamics have considered parallelization of the $O(n^2)$ [17,31] and the $O(n^3)$ algorithms [12,13,23,31]. In particular, the parallel $O(n)$ algorithms in [12,31] which are derived by parallelization of the serial $O(n^3)$ algorithms, have been shown to provide significant, though asymptotically constant, speedup over the serial $O(n)$ algorithms. However, both algorithms in [12,31] require a two-dimensional array of $O(n^2)$ processors which is different from the optimal architecture for other problems. Therefore, the results of [12,31] seem to suggest that there is no unique optimal parallel architecture for this class of problems.

In [12] it is shown that the forward dynamics also belongs to the class of NC and time and processors bound in its computation are of $O(\log^2 n)$ and $O(n^3)$, respectively. This result, that can also be extended to closed-chain case, is theoretical rather than practical since it requires a significant number of processors. In [12] it was concluded that the existence of a both time and processors optimal parallel algorithm for the problem, i.e., a parallel $O(\log n)$ algorithm with $O(n)$ processors, is an open issue. Based on the analysis of time and processors bound in the computation, it was also argued that such a parallel algorithm can be uniquely derived by parallelization of an $O(n)$ algorithm. Given the highly serial nature of the existing $O(n)$ algorithms, it was suggested that new $O(n)$ algorithms for the problem be sought with efficiency for parallel computation in mind.

Indeed, recently developed $O(n)$ algorithms for forward dynamics of both open- and closed-chain arms [7-9] validate the analysis as presented in [12]. These algorithms which are based on a completely different reformulation of the problem, are less efficient than the previous $O(n)$ algorithms for serial computation. However, they are highly parallel and their parallelization leads to both time and processors optimal parallel algorithms for the problem, i.e., to parallel $O(\log n)$ algorithms with $O(n)$ processors. The parallelization of all these new $O(n)$ algorithms is rather similar and mainly requires the parallel computation of the inverse dynamics (for evaluation of bias vector) and parallel solution of a symmetric, positive definite, block tridiagonal system. As stated before, the inverse dynamics can be computed in $O(\log n) + O(1)$ steps on a n-processors SIMD architecture with SENN interconnection. Using the cyclic reduction algorithm [20], the block tridiagonal system can also be solved in $O(\log n) + O(1)$ steps on the same architecture.

Our analysis shows that parallel computation of the considered class of problems has reached a certain level of maturity. This allows us to state two important and conclusive results regarding the parallelism in these problems as well as classification of their parallel algorithms.

1. The considered class of problems belongs to class of NC. Furthermore, they can all be solved by time optimal parallel algorithms in $O(\log n) + O(1)$ steps.

2. All the time optimal algorithms are also processors optimal. Furthermore, they can all be efficiently implemented on a unique SIMD parallel architecture with n processors interconnected through SENN structure.

B. Multi-Level Parallelism in Computation

The results of previous section seem to suggest that the best parallel architecture for this class of problem is a SIMD SENN architecture which is a rather general-purpose parallel architecture since it has been shown that it also suited for many other problems [48]. The use of such an architecture, however, may not lead to a good performance. In fact, the optimality of these parallel $O(\log n)$ algorithms should be considered in a rather asymptotic sense. Although, their speedup increases with the size of problem, these algorithms may achieve a poor speedup or even a slowdown for small size problems (see also Sec. IV). For example, for $n = 6$, the parallel $O(\log n)$ algorithm for inverse dynamics [32] achieves a speedup of 4.5. The situation is even worse for the parallel $O(\log n)$ algorithms for forward dynamics. For $n = 6$, the parallel $O(\log n)$ algorithm for the open-chain case [7] leads to a slowdown rather than speedup over the best serial $O(n)$ algorithm.

This clearly suggests that, in order to increase the speedup, further parallelism in the computation need to be exploited. Note that the parallel $O(\log n)$ algorithms discussed above exploit the parallelism at the highest computational level, i.e., at the problem level to reduce polynomial complexity. But for many problems parallelism can be exploited at several computational levels [20,21]. In this regard, the related works in [5,33,50] should be mentioned wherein special-purpose processors are designed to exploit limited parallelism at the lowest computational level.

The performance of asymptotically optimal serial and parallel algorithms reflects another salient algorithmic property of this class of problems. Unlike many other scientific computation, e.g., signal processing and linear system solution, their computational complexity, given the possible range of problem sizes, is not bounded by the size of problem but rather by the large constant coefficients in the polynomial complexity. For example, the polynomial complexity of the serial $O(n)$ algorithms for inverse and forward dynamics are given as $281n + O(1)$ and $654n + O(1)$, respectively. The polynomial complexity of the parallel $O(\log n)$ algorithms for the two problems are $63\lceil\log_2 n\rceil + O(1)$ and $1600\lceil\log_2\rceil + O(1)$, respectively ($\lceil x\rceil$ denotes the smallest integer greater than or equal to x).

This clearly indicates that the constant factor reduction is the key to achieve a significant speedup in the computation. In fact, for small problem sizes, the speedup due to the constant factor reduction can be even more significant than the speedup due to the polynomial complexity reduction. Consider the two parallel $O(n)$ algorithms for inverse [11] and forward [12] dynamics. Although they have the same asymptotic complexity as their serial counterparts, they achieve significant speedups over the serial $O(n)$ algorithms by drastically reducing the coefficient of n-dependent terms in the polynomial complexity.

The constant factor reduction can be achieved by exploitation of parallelism at lower computational levels. Obviously, a multilevel approach to exploitation of parallelism by combining both polynomial complexity and constant coefficients reduction can lead to a multiplicative effect and provide a significant overall speedup for all problem sizes. In this regard, at the problem level the algorithm's processes are computed in parallel. At the lower computational level, parallelism within each algorithm's process is exploited. The opportunity for exploitation of rather significant parallelism at lower computational levels results from the fact these parallel $O(\log n)$ algorithms have a medium to large grain size. Depending on the algorithm, each process can involve a matrix-vector operation, or a series of such oprations, or even linear system solution (for forward dynamics problem).

The motivation for an algorithmically specialized parallel architecture, rather than the SIMD SENN architecture, for this class of problem stems from the above analysis of parallelism which clearly indicates that a significant speedup can only be achieved by making architectural optimization for multilevel parallel computation.

III. Architectural Design of ASPARC

In this section, we discuss a multilevel approach to the design of ASPARC. At each level, the design is driven by the algorithmic properties of the class of problems.

1. System Level Design

The function of ASPARC is limited to the dynamic control and simulation computations. A robotic system, however, requires other functions such as path planning, trajectory generation, sensory information processing, etc.. Here, it is assumed that these functions are implemented on an External Host (EH), which can be either a single processor system, e.g., a workstation, or as is increasingly the case, a multiprocessor bus oriented system, e.g., a VME bus multiprocessor system. The first issue in the design of ASPARC is its interface to the EH.

An important algorithmic property of both dynamic control and simulation problems is that they are highly compute bound, i.e., they involve a large amount of arithmetic operations with a very limited amount of data input/ output. For an n-link robot arm and with 1 DOF per link, the data input for the dynamic control are three nx1 vectors of joint positions, velocities, and accelerations, plus a 6x1 vector of external force and moment exerted on the end-effector, and the data output are an nx1 vector of joint forces/torques. For the dynamic simulation, the data input are three $n \times 1$ vectors of joint positions, velocities, and forces/torques, plus a 6×1 vector of external force and moment, and the data output are an $n \times 1$ vector of joint accelerations. If the data input and output are sent and received by the EH, then the only communication needed between EH and ASPARC is at the beginning and at the end of a computation cycle wherein a limited amount of data needs to be exchanged.

Exploiting this property, the problem of EH/ASPARC interface is simplified. In fact, this property suggests that, at system level, the best approach is to design ASPARC as a stand-alone attached processor rather than a co-processor. The EH communicates with ASPARC through part of the bus memory which is also shared by ASPARC (Fig. 1). For interfacing to the EH, ASPARC is considered as a memory connected to the bus of EH. In the current design, this interface is considered for the VME bus which is the case for a SUN VME bus workstation or a VME bus multiprocessor system. However, this interface can easily be modified for any other type of bus by using a bus adapter. The activity of ASPARC is triggered by a function call from EH which is performed by writing a set of data in the shared memory as well as by interrupting ASPARC. The set of data includes the data input as well as the type of computation (i.e., different control schemes or simulation). The final result of the computation is sent by ASPARC to the EH through the shared memory.

2. Architecture Level Design

The architecture of ASPARC consists of two main units (Fig. 1): Internal Host (IH) and Parallel Computing Engine (PCE), which are described below.

A. Internal Host

The IH includes a general-purpose single board computer, e.g., a 68030 Motorola processor-based board with a floating-point coprocessor. The IH is the control unit of ASPARC which handles the communication with the EH and coordinates the activities of PCE. It receives the input data from the EH, interprets the function call by defining the type of computation required, distributes the input data among the processors of PCE, triggers their activities, and finally receives the results of the computation and communicates them to EH.

The IH also handles any serial and data-dependent computation required. An

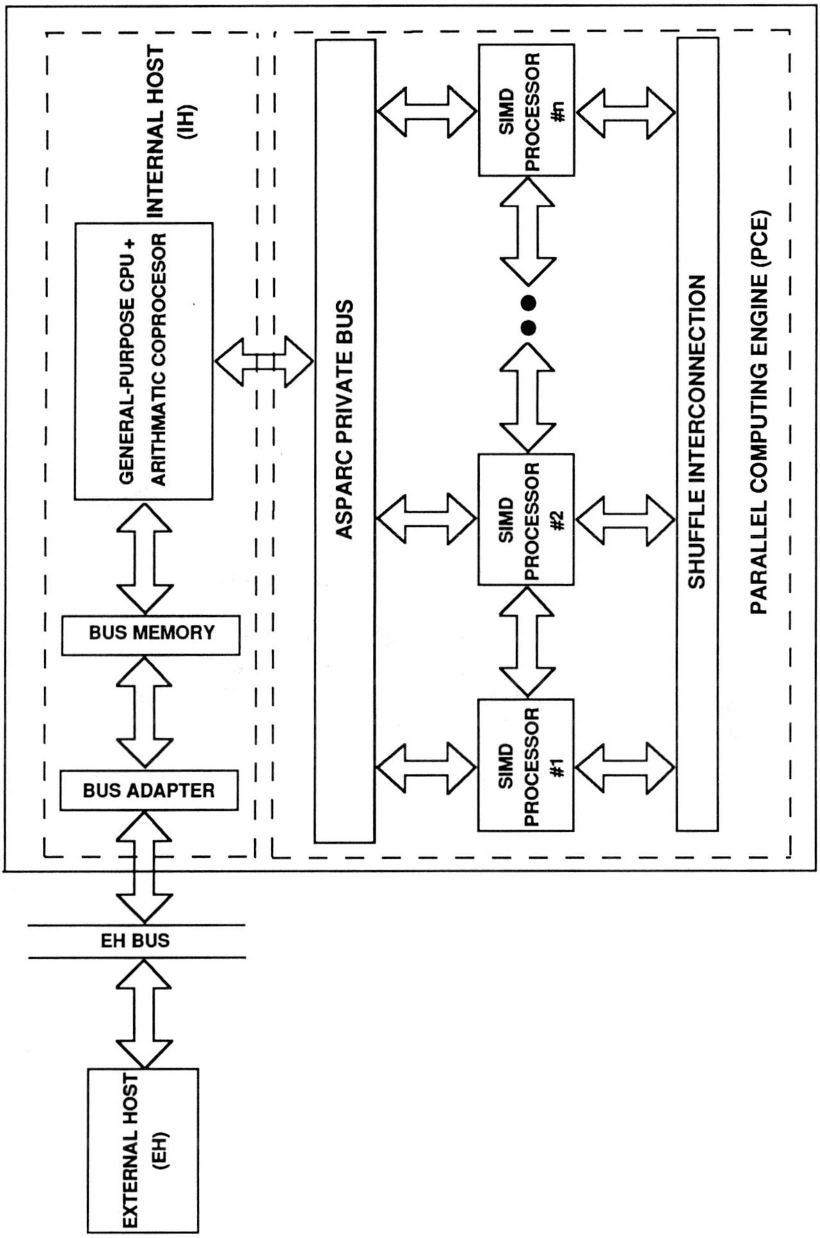

FIGURE 1. The Architecture of ASPARC

example is the inversion of Jacobian for 6 DOF spherical wrist robot arms which, as is shown in [15], can be obtained in a symbolic form. The numerical evaluation of this symbolic form is highly serial and data-dependent due to singularity, but involves a small amount of computations and hence can be efficiently performed by the IH processor by using the floating-point coprocessor. Furthermore, this computation can be overlapped with other computations required for dynamics control (see Sec. IV).

The IH communicates with the processors of PCE through the ASPARC private bus. For the IH, the processors are considered as memory banks connected to the private bus. The IH sends and receives data to and from each processor through the shared memory and triggers the processors activities by using an interrupt mechanism.

B. Parallel Computing Engine (PCE)

The heart of ASPARC is the PCE unit which exploits multilevel parallelism in different computations. The design of PCE is discussed at two levels: architecture level, wherein parallelism at problem level is exploited, and processor level, wherein parallelism at lower computational levels (matrix- vector operations) is exploited.

At the problem level, as discussed before, the optimal parallel architecture for the $O(\log n)$ algorithms (that is, for exploiting parallelism at problem level) is an n-processor SIMD architecture with SENN interconnection. This architecture can be characterized by three main features: number of processors, processors interconnection, and synchronization mechanism. In practical implementation of this architecture several issues related to these three features need to be considered.

i. Number of Processors

At the problem level maximum speedup is achieved if the parallel $O(\log n)$ algorithms are implemented with no cardinality variation, i.e., if they are implemented on n-processors architecture. Given a smaller number of processors, say $p < n$ processors, some of these algorithms, e.g., for forward kinematics, Jacobian, and inverse dynamics [32], can be implemented in $O(n/p) + O(\log n)$ steps which is optimal in terms of the number of processors. (Note, that this remains to be shown for the new algorithms for forward dynamics and operational mass matrix and its inverse.)

In order to exploit parallelism at lower computational level, each processor of PCE is designed as a stand-alone SIMD parallel architecture. Even by using the current state of the art VLSI and WSI (Wafer Scale Integration) technologies, the processors implementation is limited to one processor per board (currently, a 9U board for implementation of each processor is required). However, exploitation of maximum parallelism at problem level by using n processors (i.e., by matching the problem size) does not represent a bottleneck since, even for highly redundant robot arms, the number of processors required does not become prohibitive. But the main issue is the scalability of the architecture since the problem size can vary for different robot arms (for example, currently at JPL robot arms with 6, 7, and 8 DOFs are being used). In this regard, PCE is scalable and, depending on the problem size, additional processor boards can be added. However, the scalability of the PCE requires some modifications in the processors interconnection which is discussed below.

ii. Processors Interconnection

The optimality of the SENN interconnection for the parallel $O(\log n)$ algorithms results from the fact that it perfectly matches the communication structure inherent to these algorithms. That is, implementation of these algorithms with SENN interconnection represents a mapping with no topological variation and, therefore, leads to the minimum communication complexity of $O(\log n)$. Note that, as is shown in [14], these algorithms can also be implemented with a NN interconnection with a communication complexity of $O(n)$.

For practical implementation, however, the main issue is the scalability of the processors interconnection. The Nearest-Neighbor (NN) interconnection is scalable since adding new processors requires local modification of interconnection (i.e., adding new local communication links). But the Shuffle-Exchange (SE) interconnection is not scalable (Fig. 2 shows the SE interconnection for 8 processors). The SE interconnection is based on two routine functions as Shuffle (S) and Exchange (E) [21]. Given N processors, let $A = a_{n-1} \cdots a_1 a_0$ stand for a processor's address where $0 < A < N$ and $n = \log_2 N$. The S and E routines are given as

$$S(a_{n-1} \cdots a_{n-2} a_1 a_0) = a_{n-2} \cdots a_1 a_0 a_{n-1}$$

$$E(a_{n-1} \cdots a_1 a_0) = a_{n-1} a_1 \bar{a}_0$$

The E routine can be implemented by using NN interconnection. However, the interconnection required for the implementation of S routine is not scalable. That is, if the number of processors is increased then a global modification of processors interconnection is required. As stated before, the implementation of each SIMD processor requires a complete board and, therefore, interconnection for S routine needs to be implemented as a board- to-board connection. Given the limited number of boards, even for redundant robot arms, the implementation and modification of this interconnection does not represent a major bottleneck.

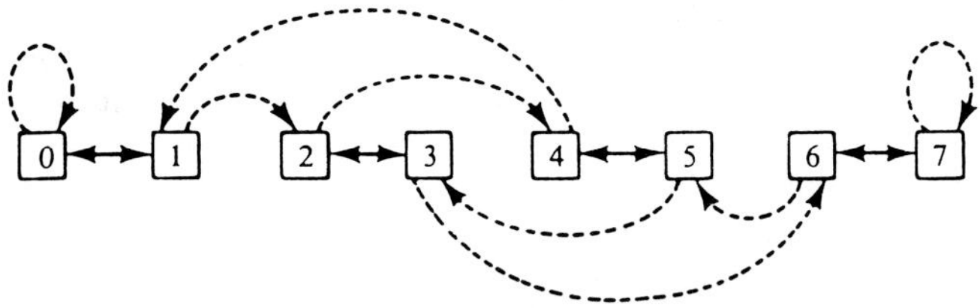

Figure 2 Shuffle-Exchange Recirculating Network for N = 8 (From[21]). Solid lines are Exchange and Dashed lines are Shuffle.

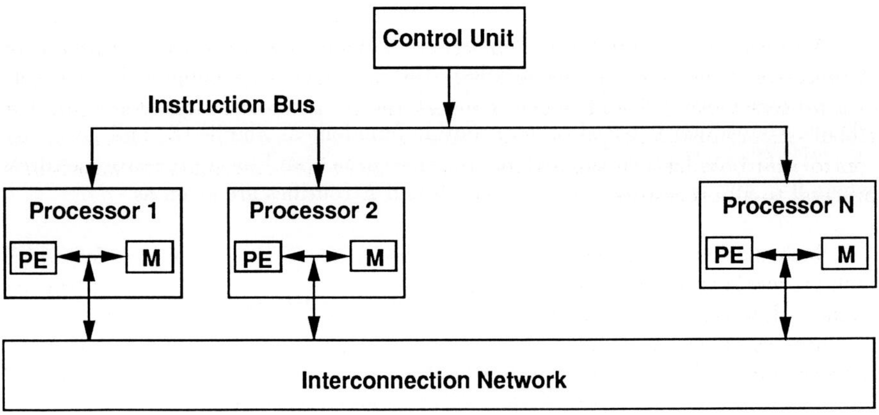

Figure 3 A Typical SIMD Parallel Architecture

The processors interconnection of PCE is shown in Fig. 1. In addition to NN and S interconnections the ASPARC private bus can also be used as a global bus for communication among SIMD processors. A ring topology is also obtained by using the NN interconnection and the private bus as the direct communication link between SIMD 1 and SIMD n. The ring topology in connection with the MIMD-SIMD synchronization mode (see below) is particularly suitable for many computations, e.g., parallel computation of inverse dynamics [11].

In addition to this extensive processors interconnection, a fast communication scheme is also employed to further reduce communication overhead. The communication among processors can be performed through shared memories (register files) with a speed of one datum per clock without any communication initiation overhead.

iii. Synchronization Mechanism

The optimality of SIMD synchronization mechanism for the parallel $O(\log n)$ algorithms results from the regularity and determinacy (data-independency) of their main operations. However, due to the medium to coarse grain size of the algorithms, they can also be implemented with a MIMD synchronization mechanism though with less efficiency.

A typical SIMD architecture consists of a single global control unit and a set of processors wherein each processor has a simple internal architecture and consists of a processor element and a local memory (Fig. 3). In such an architecture the global synchronization of processors activities is performed at instruction level: the control unit broadcasts a same global instruction (which represents the algorithm's process) to all processors.

Such mechanism is particularly suitable for fine grain algorithms for which each process represents a rather primitive operation, e.g., multiply, add, or transfer scalars. However, the processes of the parallel $O(\log n)$ algorithms considered here are more complex and involve matrix-vector operations or a series of such operations (including linear system solution). Therefore, for these algorithms each global instruction involves a rather complex set of primitive operations.

Coordination of the activities of the processor elements at the primitive operation level by global control unit requires a highly complex structure for the global control unit and its instructions. This complexity can be reduced by designing a local control unit for each processor element. The function of the local control unit is to interpret the global instructions sent by the global control unit, to properly sequence the set of primitive operations, and to fetch and store the data from and to the local memory. However, the local control unit should also provide further functionalities. As stated before, in order to exploit parallelism at lower computational level (i.e., within each process), each processor of PCE is a SIMD parallel architecture and as a result the control unit of each processor has a rather complex structure.

Although, the processes of the parallel $O(\log n)$ algorithms are complex, which dictates the need for more complex local control units, these algorithms require few global synchronization, which can be exploited to simplify the structure of the global control unit. The computation of these algorithms consists of few parts where each part is performed in either $O(1)$ step - which implies that the operations of the SIMD processors are carried out completely in parallel without any need for global synchronization - or in $O(\log n)$ steps - which implies that the global synchronization takes place $O(\log n)$ times. If all processors are driven by the same clock then, due to the regularity and determinacy of the algorithms' processes, a clock-based global synchronization mechanism similar to systolic arrays [28] can be employed. Using such a mechanism eliminates the need for having a global control unit.

Note that, compared to the classical systolic arrays, each processor of PCE instead of being a simple processor is a SIMD parallel architecture. In this regard, the PCE is a two-level SIMD-SIMD architecture. On the other hand, since each processor is a stand-alone SIMD architecture with its own control unit, the processors can also perform their operations in parallel and asynchronously. In this mode, the PCE functions as a two-level MIMD-SIMD architecture wherein the synchronization among SIMD processors is achieved by using a data-driven mechanism similar to wave front arrays [29]. This MIMD-SIMD mode is particularly suitable for a different approach to parallel computation of inverse dynamics [11] which allows the communication and computation to be overlapped.

C. SIMD Processor Design

The main challenge in the design of ASPARC is the architecture of SIMD parallel processors. While at the problem level the $O(\log n)$ algorithms exhibit similar structural properties which allow their implementation on a unique architecture, at lower computational levels (or, within each algorithm's process) they involve different computations with different properties. The parallel algorithms for forward kinematics and Jacobian, and inverse dynamics involve a series of operations on 3×3 matrices and 3×1 vectors, and the parallel algorithms for open- and closed-chain forward dynamics involve operations on 5×5, 5×6, and 6×6 matrices and 5×1 and 6×1 vectors as well as the solution of symmetric, positive definite, linear systems where matrices are of 5×5 (for joints with 1 DOF).

Here, the exploitation of parallelism is not constrained by algorithmic design (it is rather straightforward) but by the degree to which the required architectural complexity can be efficiently managed. It is impossible to design an architecture that can exploit maximum parallelism in all these computations and achieve the minimum computation time in terms of number of operations. In fact, even for a specific algorithm, say forward kinematics and Jacobian which involves simple operations, exploitation of maximum parallelism requires very complex architecture, making its implementation impractical (see Sec. IV).

Our design strategy is based on efficient tradeoffs to exploit parallelism as much as possible for all these different computations while limiting the architectural complexity. In order to efficiently exploit parallelism in a wide variety of operations, the architecture should provide a high degree of flexibility and reconfigurability. These two features are incorporated by using an elaborated memory hierarchy, an extensive data path, and a distributed control unit. Figure 4 shows the architecture of the SIMD processor. In the following some architectural aspects of the SIMD processor are discussed.

i. Arithmetic Unit

The key choice that significantly affects the design of SIMD processor is the degree to which the parallelism in various operations is exploited. This is determined by the number of PEs. It might seem that by increasing the number of PEs the exploitable parallelism is also increased. Practically, however, the number of PEs is limited by the memory bandwidth. For the PEs to be used efficiently, the data should be fetched and stored from and to the memory as well as routed among PEs with adequate speed. The choice of number of PEs is based on the size of matrices and vectors, the type of operations, and the complexity of required memory organization and data path.

The arithmetic unit includes 6 PEs and 2 Table Lookup Units (TLU). Depending on the type of operation, the PEs can be grouped in different ways. They can work as a single group of 6 coupled PEs to perform parallel computation of matrix-vector operations or they can work as 6 independent PEs and perform 6 scalar operations in parallel. They can also work as two independent groups of 3 coupled PEs for parallel computation of matrix-vector operations where matrices are 3×3 and vectors are 3×1. In this case, two such operations can be performed in parallel.

The extensive data path and the Interconnection Switch (IS1) allow a variety of interconnection topologies among PEs. The 6 PEs can be interconnected as a linear array (which is suitable for linear system solution) or as a ring (Figs. 5a and 5b). They can also be interconnected as two independent linear arrays of 3 PEs or as two rings of 3 PEs (Figs. 5c and 5d). Using the NN interconnection among SIMD processors and exploiting the facts that the SIMD processors are driven with the same global clock and that the communication between adjacent SIMD processors can be performed with a speed of one datum per clock, the two topologies of Figs. 5a and 5c can be extended to include the PEs of all SIMD processors. In this case,

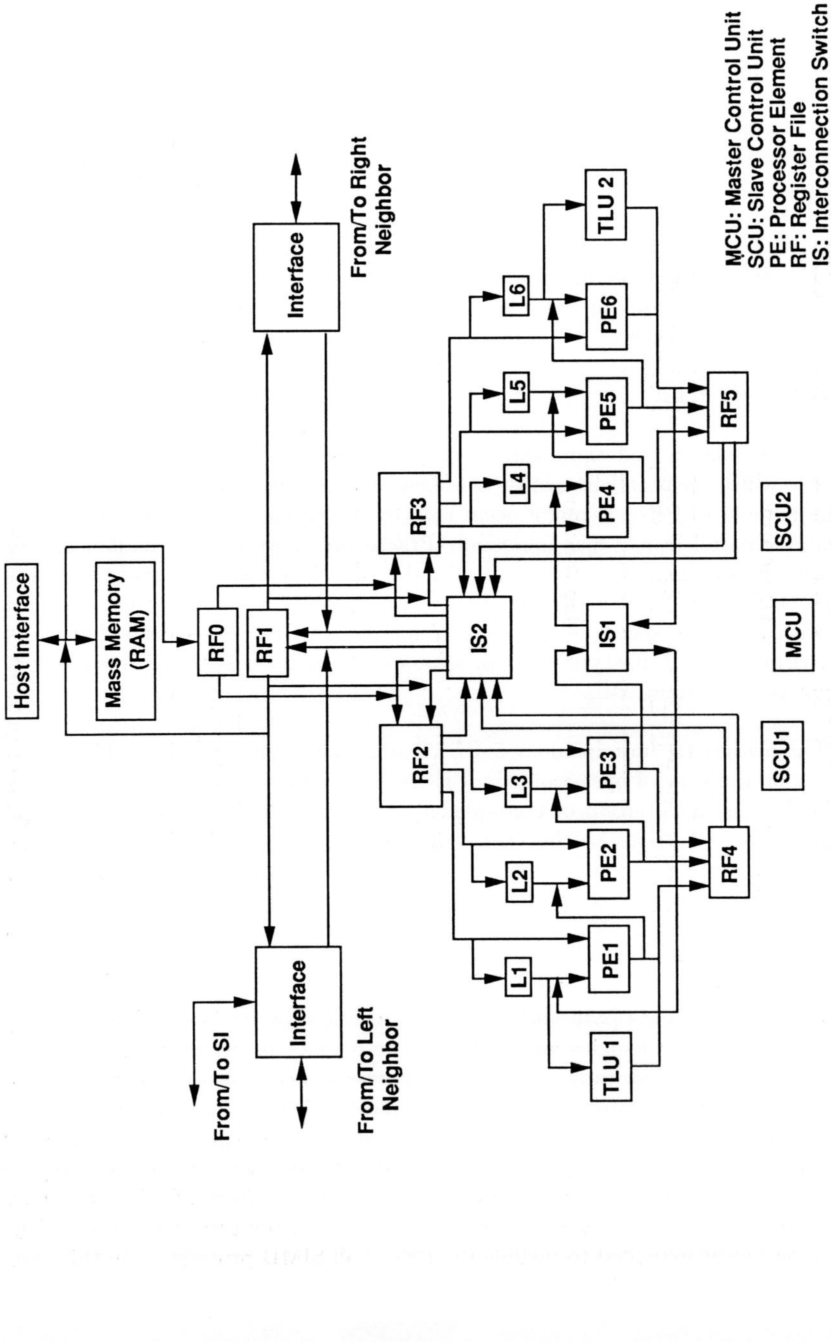

Figure 4 SIMD Parallel Processor Architecture

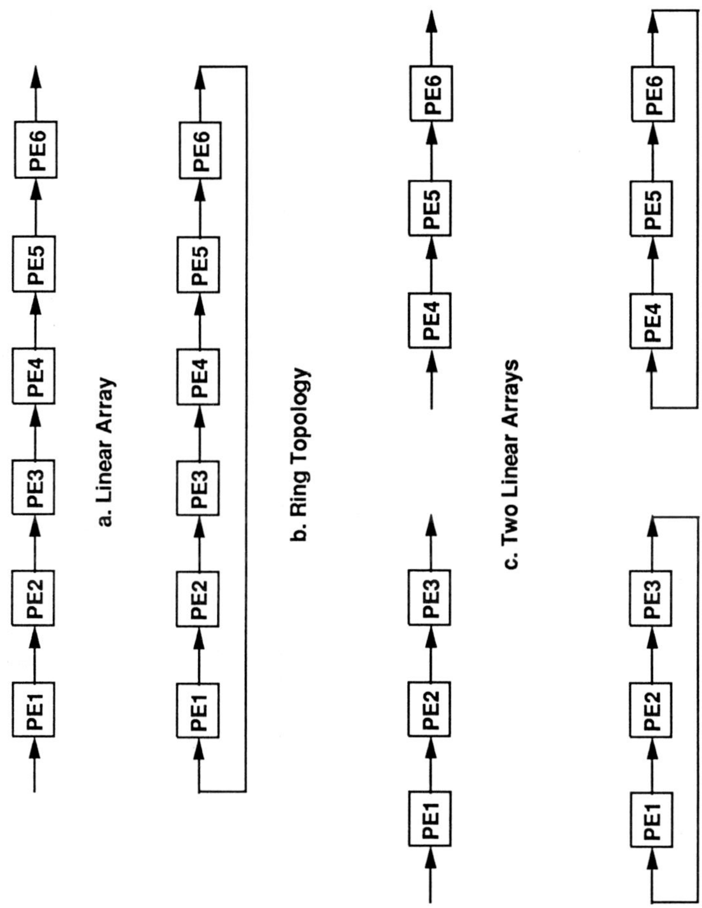

Figure 5 Different Configurations of Processor Elements

the PCE can be configured as a long linear array of 6n PEs or as a two independent linear arrays of 3n PEs. In these two configurations, the PCE can operate similar to systolic arrays in the sense that all PEs are globally synchronized by a same clock and that the computation and communication can be overlapped.

The current design of ASPARC is mainly based on the use of commercially available VLSI components. Although it is tried to keep the architectural design independent of any specific component, a judicious choice of the components can reduce both hardware and programming complexity. This is particularly the case for the choice of PEs. There are many commercially available floating-point processors (or, DSP chips) which offer a high computing performance and can be used as PE (see for example [53]). However, most of these chips have a pipelined architecture and require rather complex control instructions. Note that, due to the small dimension of matrices and vectors, the pipeline capability of these chips, while increases the programming complexity, does not offer any speedup in matrix-vector operations.

A suitable choice for PEs is a simple floating-point processor such as AMD Am29325 32-bit floating-point processor. This processor has a flow-through (non pipelined) architecture [52] and performs the different floating-point operations in one clock cycle. The multiply and add operation, which is required for sum-of-product operations, e.g., for vector-dot operation, is internally supported and takes two clock cycles. It also has a simple control instruction (4-bit instruction). These two features are particularly suitable since they allow integration of 6 PEs within each SIMD processors without significantly increasing the hardware and programming complexity. Like many other DSP chips, this processor does not directly support division operation (which is required for linear system solution). This operation is performed by using the Newton-Raphson method which is iterative and requires a seed value to start the iterations. The two TLU's are incorporated to provide the seed values required for division operations.

ii. Memory Organization and Data Paths

A classical bottleneck in the design of any computer architecture (particularly parallel architectures) is the gap between the speed of memory and that of CPU. Most often, the computation speed is not bounded by the speed of CPU but by the memory bandwidth which limits the speed by which the data can be loaded and stored. For the SIMD processor, given the use of 6 fast PEs, each performing one operation per clock cycle, as computing units, the memory bandwidth is even more critical.

Two other issues need to be considered in the design of memory structure and data paths. The first is that parallel computation of most of the matrix-vector operations by the SIMD processor requires that the same data be available for several PEs. The second is that some data generated during an operation are reused in computation of other operations. Therefore, in addition to high bandwidth, both memory structure and data path should provide a high degree of flexibility and reconfigurability for efficient data routing among PEs.

However, the computations performed by each SIMD processor involve a limited set of both constant and variable data. As a result of process-to- processor allocation of the O(log n) algorithms, each SIMD processor performs the computation associated with one link. Therefore, it needs to store a rather limited set of constant kinematic and dynamic parameters of its associated link. It also generates a limited set of kinematic and dynamic variables of its associated link. This implies that each SIMD processor requires a rather small size memory. Exploiting this property, two other techniques are employed to achieve both high memory bandwidth and flexible data routing capability while limiting the hardware complexity.

Given the small size of required memory, fast small memories (a set of register files) are used in the form of a distributed memory structure. The SIMD processor is run with two clocks: major clock and minor clock. The minor clock is twice faster than the major clock. The PEs are run with the major clock, i.e., each floating-point operations takes one major clock cycle, while data movement, i.e., read/write and input/output, takes one minor clock cycle. This time multiplexing increases the memory bandwidth and allows a high degree of overlapping between computation and data movement.

In addition to the time multiplexing, a multilevel hierarchical memory structure is designed to provide the required high bandwidth and flexibility (see Fig. 4). The mass memory and the Register File RF0 are both part of ASPARC private bus memory and are shared with the IH. The mass memory is a small (1K) RAM memory and contains the link constant parameters which are loaded by the IH prior the start of computation. RF0 is a 3 ports (1 input and 2 outputs) register file which is used for fast communication of input data with the IH. For each instant of the computation, the corresponding constant parameters are first moved to RF0. Any data stored in RF0 can be then made available simultaneously to all PEs through RF2 and RF3.

RF1 is a four ports (two input ports and two output ports) register file which is also part of ASPARC private bus memory and is used for fast communication of final data to IH. It is also used for communicating (both sending and receiving) data to and from adjacent SIMD processors or through SI. Again any data stored in RF1 can be made available simultaneously to all PEs through RF2 and RF3.

At each major clock cycle 12 operands need to be fetched for PEs. This high bandwidth requirement is met by using time multiplexing and the two register files RF2 and RF3 which are six ports (two input ports and four output ports) register files and provide the operands for PEs as well as TLUs. At each minor clock cycles four data can be read from and two data can be written in RF2 and RF3. RF2 provides the operands for the group of PEs consisting of PE1, PE2, and PE3 as well as TLU1. RF3 provides the operands for the group of PEs consisting of PE4, PE5, and PE6 as well as TLU2. For performing floating-point operations, at each minor clock cycle six operands are read from RF2 and RF3. These operands are then latched by using the Latches (L) and remain available at the input port of PEs. At the next minor clock cycle, six other operands are read from RF2 and RF3. Therefore, within two minor clock cycles (or, one major clock cycle) all the operands are available at the input ports of PEs. The data movement by RF2 and RF3 and the floating-point operations by PEs are overlapped. That is, while PEs are performing the floating-point operations, the next set of operands are fetched by RF2 and RF3, thus allowing the PEs to work with the maximum speed without any wait cycle for fetching the data. The direct data movement between RF2 and RF3 can be also performed through IS2.

RF4 and RF5 are five ports (three input ports and two output ports) register files. They are used for storing and routing the results of computations performed by PEs as well as the results of table look-up operations read from TLUs. At every minor clock cycle, the results of all PEs' operations can be written simultaneously into RF4 and RF5. By using IS2, these data can then be written into RF2 and RF3 (four data per minor clock), or to RF1 (two data per minor clock cycles), or directly communicated to other SIMD processors (one datum per minor clock cycle). The data movements to and from RF4 and RF5 are overlapped with the PEs operations as well as the loading data from RF2 and RF3.

It should be emphasized that, like PEs, the current design of SIMD processors are based on the use of commercially available register files. Also, commercially available components are used for the two ISs. However, future implementations of the SIMD processors can become more area efficient by using VLSI and particularly WSI technologies for the implementation of special-purpose register files and ISs.

iii. Control Unit(s)

ASPARC is an application-specific architecture targeted for a limited set of well defined problems. The parallel O(log n) algorithms considered here are general in the sense that they are valid for any robot arm since they are based on recursive numerical methods. Each SIMD processor also performs a set of limited and well defined computations which are the same regardless of any considered robot arm. SIMD processors are microprogrammable and do not support high level language. In fact, given the complexity of SIMD processors, developing a compiler is a formidable task and more complex and time consuming than the development of microcodes for the target computations!

Each SIMD processor is a VLIW (Very Long Instruction Word) processor which in each clock cycle (both major and minor) performs many operations in parallel. In order to reduce the complexity and the size of instruction word, a distributed control structure consisting of one Master Control Unit (MCU) and two Slave Control Units (SCU) is designed. The SCUs are run with the minor clock. Each SCU controls the activities of a group of 3 PEs which include fetching the operands, issuing the instruction for PEs, and routing the results of the computations performed by the PEs. The two SCUs can operate independently. This is particularly suitable for microcode development since for some operations, only one group of PEs is used and hence only one SCU needs to be active and programmed. The MCU is run with the major clock cycle and coordinate the activities of the two SCUs, e.g., by activating one or both SCUs for performing the computation. It also determines the configuration of PEs and data routing paths by controlling the two ISs.

IV. Implementation of Parallel Algorithms on ASPARC

In this section two implementation examples of parallel algorithms on ASPARC are discussed. The first is the parallel algorithm for computation of forward kinematics and Jacobian. This algorithm is conceptually simple and is used to better illustrate the significance of parallelism in matrix-vector operations as well as the way by which these operations are performed on SIMD processors. The second example is the parallel implementation of RAC scheme which is used to illustrate the sequencing of different computations as well as the way that synergism can be exploited in different computations.

A. Parallel Computation of Forward Kinematics and Jacobian

On serial processors, a conventional approach for computation of forward kinematics and Jacobian is based on the use of symbolic methods. In these methods, first the symbolic expressions are derived and then, for each set of new data input (joint positions), the numerical values of the expressions are evaluated. While for simple 6 DOF robot arms, e.g., PUMA arm, the symbolic expressions are rather simple, for more complex redundant arm, e.g., 7 DOF Robotic Research arm, the symbolic expressions (particularly for Jacobian) are very complex and their numerical evaluation becomes very time consuming. For parallel computation, recursive methods are more efficient than symbolic methods since the computation of symbolic expressions is highly serial.

An efficient recursive algorithm which combines the computation of both forward kinematics and Jacobian (${}^1J_E$) wherein the superscript denotes the frame in which vectors and tensors are described (see also Fig. 6, Table I, and Appendix).

For $i = 1$ to n, Do

$$O(1, i+1) = O(1, i)O(i, i+1) \quad \text{with} \quad O(1, 1,) = U \text{ (Unit matrix)} \qquad (1)$$

$Q(i), \dot{Q}(i), \ddot{Q}(i)$: Position, velocity, and acceleration of joint i.

$Q, \dot{Q}, \ddot{Q}$: n x 1 vector of joint positions, velocities, and accelerations.

$\tau(i)$: Joint i control force/torque.

τ : n x 1 vector of joint forces/torques.

$m(i)$: Mass of link i

$I(i)$: Second moment of mass of link i about its center of mass.

$P(j,i)$: Position vector frame point i to point $j, P(i+1,i) = P(i)$.

$O(j,i)$: 3x3 matrix describing orientation of frame i with respect to frame j.

$\omega(i+1)$: Angular velocity of link i (frame $i+1$).

$V(i+1)$: Linear velocity of link i (origin of frame $i+1$).

$\dot{\omega}(i+1)$: Angular acceleration of link i (frame $i+1$).

$\dot{V}(i+1)$: Linear acceleration of link i (origin of frame $i+1$).

$\dot{V}_c(i+1)$: Linear acceleration of center of mass of link i.

$\dot{\omega}'$ (i+1): Angular acceleration of link i (frame $i+1$) obtained by setting $\ddot{Q} = 0$.

$\dot{V}'(i+1)$: Linear acceleration of link i (origin of frame $i+1$) obtained by setting $\ddot{Q} = 0$.

$F(i)$: Force exerted on center of mass of link i.

$N(i)$: Moment exerted on center of mass of link i.

$f(i)$: Force exerted on link i by link $i-1$.

$n(i)$: Moment exerted on link i by link $i-1$.

Table I. Link Kinematic and Dynamic Parameters

m: Multiplication.

a: Addition.

SV: Scalar-Vector Multiplication.

VA: Vector Addition.

VC: Vector-Cross Product.

VD: Vector-Dot Product.

MV: Matrix-Vector Multiplication.

MM: Matrix-Matrix Multiplication.

Table II. Notation used in Computational Complexity Analysis

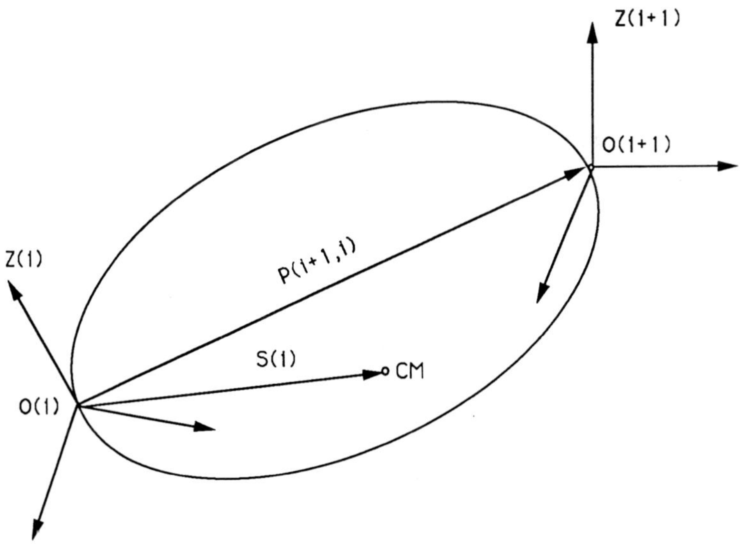

Figure 6 Link, Frames, and Position Vectors
CM: Center of Mass

End-Do

For $i = 1$ to n, Do

$$P(i) = O(1, i+1)^{i+1} P(i) \tag{2}$$

End - Do

For $i = n$ to 1

$$^1P(n+1, i) = \ ^1P(n+1, i+1) + \ ^1P(i) \quad \text{with } ^1P(n+1, n+1) = 0 \tag{3}$$

End-Do

For $i = 1$ to n, Do

$$J_{ri} = \ ^1Z(i) = O(1, i)_{(3)} \quad \text{(third column of } O(1, i)) \tag{4a}$$

$$J_{ti} = \ ^1Z(i) \times \ ^1P(n+1, i) \tag{4b}$$

End-Do

On a serial processor, the computation cost of this algorithm is obtained from Eq. (1)-(4b) as $(42m+30a)n$ - $(27m+21a)$.

With n processors, the computation of First-Order Linear Recurrences (FOLR) in Eqs. (1) and (3) can be performed in $\lceil \log_2 n \rceil MM$ and $\lceil \log_2 n \rceil VA$, respectively. The computations in Eqs. (2), (4b) require $1MV$ and $1VC$, respectively, and Eq. (4a) does not require any computation. If parallelism in matrix-vector operations is not exploited, i.e., only parallelism at problem level is exploited, then the cost of parallel computation of the algorithm is obtained as $(27m + 21a)\lceil \log_2 n \rceil + (15m + 9a)$. Assuming the same time for multiplication and addition (i.e., $m = a$), for $n = 6$ (e.g., for PUMA arm) and $n = 8$ (e.g., for AAI 8-DOF robot arm), this represents a speedup of about 2.3 and 2.5, respectively.

It is instructive to analyze the minimum computation time (in terms of minimum number of operations) and hence the absolute maximum speedup that can be achieved for this algorithm. If maximum parallelism in matrix-vector operations is exploited, then the cost of different operations is given as

$$MM = 1m + 2a, MV = 1m + 2a, VC = 1m + 1a, \text{ and } VA = 1a.$$

If maximum parallelism at both problem level and matrix-vector operations level is exploited, then the cost of parallel computation of the algorithm is given as $(1m + 3a)\lceil \log_n \rceil + (2m + 3a)$. For $n = 6$ and $n = 8$ (assuming $m = a$), this represents significant speedup of about 25 and 22, respectively.

This clearly indicates that, theoretically, for this algorithm the parallelism in matrix-vector operations is more significant than the parallelism at problem level. Practically, however, exploitation of maximum parallelism in some matrix-vector operations requires very complex architecture. While for VA and VC operations maximum parallelism can be exploited by using 3 and 6 PE's with a simple PE's interconnection, for MV and MM operations this requires 18 and 27 PE's with a complex PE's interconnection.

For the parallel computation of forward kinematics and Jacobian on the ASPARC, it is assumed that the constant parameters of link i as well as the data input (joint position) for joint i already reside in the memory of the ith SIMD Parallel Processor. The parallel algorithm, along with process-to- processor allocation is given below where PR_i stand for the ith SIMD Parallel Processor.

Step 1:

For $\eta = 1$ to $\lceil \log_2 n \rceil$, Do

For $i = 2$ to $n+1$, Do in Parallel

$$\begin{aligned} & PR_{i-1} : O(i-2^\eta, i) = O(1,i) \\ & \qquad 1 + 2^\eta > 1 + 2^{\eta-1} \geq i \\ & PR_{i-1} : O(i-2^\eta, i) = O(1,i) = O(1, i-2^{\eta-1}) O(i-2^{\eta-1}, i) \\ & \qquad 1 + 2^\eta \geq i > 1 + 2^{\eta-1} \\ & PR_{i-1} : O(i-2^\eta, i) = O(i-2^\eta, i-2^{\eta-1}) O(i-2^{\eta-1}, i) \\ & \qquad i > 1 + 2^\eta > 1 + 2^{\eta-1} \end{aligned} \tag{5}$$

End - DO Parallel

End - Do

Step 2:

For $i = 1$ to n, Do in Parallel

$$PR_i : \ {}^1P(i) = O(1, i+1)^{i+1}P(i) \tag{6}$$

End - Do Parallel

Step 3:

For $\eta = 1$ to $\lceil \log_n \rceil$, Do

For $i = 1$ to n, Do in Parallel

$$\begin{aligned} &PR_i : P(i+2^{\eta}, i) = P(n+1, i) \\ &\quad i \geq n+1-2^{\eta} > n+1-2^{\eta-1} \\ &PR_i : P(i+2^{\eta}, i) = P(n+1, i) = P(n+1, i+2^{\eta-1}) + P(i+2^{\eta-1}, i) \\ &\quad n+1-2^{\eta} \geq i > n+1-2^{\eta-1} \\ &PR_i : P(i+2^{\eta}, i) = P(i+2^{\eta}, i+2^{\eta-1}) + P(i+2^{\eta-1}, i) \\ &\quad n+1-2^{\eta} > n+1-2^{\eta-1} > i \end{aligned} \tag{7}$$

End-DO Parallel

End-Do

Step 4:

For $i = 2$ to $n+1$, Do in Parallel

$$PR_{i-1} : \ {}^1Z(i) = O(1,i)_{(3)} \ \text{(third column of } O(1,i)) \tag{8}$$

$$PR_{i-1} : \ \text{Send } {}^1Z(i) = J_{ri} \text{ to } PR_i \tag{9}$$

$$PR_{i-1} : J_{li-1} = {}^1Z(i-1) \times {}^1P(n+1, i-1) \tag{10}$$

End-Do Parallel

Note that the terminating conditions in the first and second Eqs. of (5) and (7) are checked by the SIMD processor's control unit.

The matrix-vector operations are computed by each SIMD processor as follows. The MV operation in Eq. (6) is decomposed into 3 VD operations which are performed by 3 PE's in parallel, leading to a cost of $3m + 2a$. The MM operation in Eq. (5) is decomposed into $3MV$ operations and is performed in two steps. In the first step, $2MV$ operations are performed in parallel by 6 PE's and in the second step the third MV operation is performed in parallel by 3 PE's, leading to a cost of $6m + 4a$ for MM operation. The VA operation in Eq. (7) is performed by 3 PE's in parallel with a cost of $1a$. The VC operation in Eq. (10) can be performed by 3 PE's in parallel with a cost of $2m + 1a$. However, it can be performed faster by using 6 PE's with a cost of $1m + 1a$, which illustrates an efficient use of PE's interconnection.

The cost of parallel computation of forward kinematics and Jacobian on ASPARC is then obtained as $(6m + 5a)\lceil log_2 n\rceil + (4m + 3a)$. For $n = 6$ and $n = 8$, this represents a speedup of 9.6 and 10.7, respectively, which shows significant improvement over the rather poor speedups (2.3 and 2.5) achievable by exploitation of parallelism at problem level only.

B. Parallel Implementation of Resolved Acceleration Control (RAC)

The RAC scheme [34] achieves decoupling and linearization of the End-Effector (EE) motion as follows. Let the 6x1 vectors $X, \dot{X}$, and $\ddot{X}$ denote the configuration (position and orientation), spatial (linear and angular) velocity, and spatial (linear and angular) acceleration of the EE's frame, i.e., frame $n + 1$. Also, let subscripts a, d, and c denote the actual, desired, and commanded value of the vectors. Given a desired trajectory for the EE motion (which specifies $X_d, \dot{X}_d$, and $\ddot{X}_d$ in advance), the commanded acceleration of the $EE, \ddot{X}_c$ is computed as

$$\ddot{X}_c = \ddot{X}_d + k_1(\dot{X}_d - \dot{X}_a) + k_2(X_d - X_a) \tag{11}$$

where k_1 and k_2 are given constant. The spatial velocity and acceleration of the EE are given as

$$\dot{X} = J_E\dot{Q} \tag{12}$$

$$\ddot{X} = J_E\ddot{Q} + \dot{J}_E\dot{Q} \tag{13}$$

From Eqs. (11) and (13), the vector of joint accelerations is computed as

$$\ddot{Q} = J_E^{-1}(\ddot{X}_c - \dot{J}_E\dot{Q}) \tag{14}$$

and finally the vector of joint control forces/torques, τ, is obtained by computing the Newton-Euler (N-E) algorithm [35] as

$$\tau = f(Q, \dot{Q}, \ddot{Q}, F_E)$$

where F_E is the 6x1 vector of spatial (force and moment) force exerted on EE.

As can be seen, the implementation of the RAC requires the computation of forward kinematics (to obtain X_a), forward velocity propagation (to obtain $\dot{X}_a$), the term $\dot{J}_E\dot{Q}$, Jacobian and its inverse (i.e., solution of the linear system in Eq. (14)), and inverse dynamics (the N-E algorithm). In this regard, the implementation of RAC is not only computation-intensive but also represents an illuminating example for the different kinematic and dynamic computations that need to be combined. Despite the validity of the RAC as an advanced task space control scheme, there has not been any report on a unified approach for its parallel implementation.

One seemingly misunderstood issue regarding the implementation of RAC is the computation of $\dot{J}_E$ (or $\dot{J}_E\dot{Q}$). In [34] the implementation of RAC is considered for the 6-DOF Stanford arm and there is no specification on the technique for the computation of $\dot{J}_E$ or $\dot{J}_E\dot{Q}$. Of course, if J_E is obtained in a symbolic form then the symbolic expression for $\dot{J}_E$ can be also derived by simple differentiation. However, the symbolic expression of $\dot{J}_E$ (particularly for redundant arms) can be very complex, making its numerical evaluation time consuming. There have been several reports on the development of numerical methods for the computation of $\dot{J}_E$ (see for example [1]).

However, as can be seen from Eq. (14), the explicit computation of $\dot{J}_E$ is not needed and only the term $\dot{J}_E\dot{Q}$ needs to be computed explicitly. The misunderstanding regarding the computation of $\dot{J}_E$ seems to result from the lack of attention to the physical interpretation of the term $\dot{J}_E\dot{Q}$. To see this, note that if in Eq. (13) the vector of joint accelerations ($\ddot{Q}$) is set to zero then the vector

$$\ddot{X}_a' = \dot{J}_E\dot{Q} \tag{15}$$

represents the actual spatial acceleration of the EE due to the joint velocities, that is, the Coriolis and centrifugal accelerations of the EE frame which are velocity dependent. Therefore, $\ddot{X}_a'$ can be directly computed by using the recursive propagation of the accelerations in the N-E algorithm and setting the vector of joint accelerations to zero (see Table III). The greater efficiency of this method results from the fact that not only the explicit computation of $\dot{J}_E$ and the matrix-vector multiplication for computing $\dot{J}_E\dot{Q}$ are avoided but also the synergism in computing the N-E algorithm and the vector $\ddot{X}_a'$ is exploited. In the Appendix, the explicit expression of $\dot{J}_E$ is derived and it is shown that the matrix-vector multiplication of $\dot{J}_E\dot{Q}$ leads to recursive computations which exactly correspond to the recursive propagation of accelerations in N-E algorithm while setting $\ddot{Q}$ to zero.

For the parallel implementation of RAC, the results of the computation of forward kinematics and Jacobian are needed. It is assumed the computation of parallel algorithm described in previous subsection is already completed and that the results reside in the SIMD processors. It is also assumed that the linear system solution in Eq. (14) is performed by IH or some other processors. A serial algorithm

Step I: Propagate Angular and Linear Velocities

${}^1V(1)$ and ${}^1\omega(1)$ are known initial conditions.

For $i = 1$ to n

$$ {}^1\omega(i+1) = {}^1\omega(i) + {}^1Z(i)\dot{Q}(i) \quad (16)$$

$$ {}^1V(i+1) = {}^1V(i) + {}^1\omega(i+1) \times {}^1P(i) \quad (17)$$

$$ {}^1\dot{\omega}'(i+1) = {}^1\dot{\omega}'(i) + {}^1\omega(i) \times {}^1Z(i)\dot{Q}(i) \quad (18)$$

$$ {}^1\dot{V}'(i+1) = {}^1\dot{V}'(i) + {}^1\dot{\omega}'(i+1) \times {}^1P(i) + {}^1\omega(i+1) \times ({}^1\omega(i+1) \times {}^1P(i)) \quad (19)$$

Step II. Propagate Angular and Linear Accelerations

${}^1\dot{V}(1)$ and ${}^1\dot{\omega}(1)$ are known initial conditions.

For $i = 1$ to n

$$ {}^1\dot{\omega}(i+1) = {}^1\dot{\omega}(i) + {}^1Z(i)Q(i) + {}^1\dot{\omega}'(i+1) \quad (20)$$

$$ {}^1\dot{V}(i+1) = {}^1\dot{V}(i) + {}^1\dot{\omega}(i+1) \times {}^1P(i) + {}^1\dot{V}'(i+1) \quad (21)$$

$$ {}^1S(i) = 0(1, i+1)^{i+1}S(i) \quad (22)$$

$$ {}^1\dot{V}_c(i) = {}^1\dot{V}(i) + {}^1\dot{\omega}(i+1) \times {}^1S(i) + {}^1\omega(i+1) \times ({}^1\omega(i+1) \times {}^1S(i)) \quad (23)$$

Step III. Compute Joint Forces/Torques

${}^1f(n+1)$ and ${}^1n(n+1)$ are known initial conditions.

For $i = n$ to 1

$$ {}^1F(i) = m(i)\,{}^1\dot{V}_c(i) \quad (24)$$

$$ {}^{i+1}\omega(i+1) = O(i+1,1)\,{}^1\omega(i+1) \text{ and } {}^{i+1}\dot{\omega}(i+1) = O(i+1,1)\,{}^1\dot{\omega}(i+1) \quad (25)$$

$$ {}^{i+1}N(i) = {}^{i+1}I(i)\,{}^{i+1}\dot{\omega}(i+1) + {}^{i+1}\omega(i+1) \times ({}^{i+1}I(i)\,{}^{i+1}\omega(i+1)) \quad (26)$$

$$ {}^1N(i) = O(1, i+1)^{i+1}N(i) \quad (27)$$

$$ {}^1f(i) = {}^1F(i) + {}^1f(i+1) \quad (28)$$

$$ {}^1n(i) = {}^1N(i) + {}^1S(i) \times {}^1F(i) + {}^1n(i+1) + {}^1P(i) \times {}^1f(i+1) \quad (29)$$

$$ \tau(i) = {}^1Z(i).\,{}^1n(i) \quad (30)$$

Table III. Modified Newton-Euler Algorithm

is developed (Table III) by modifying the N-E algorithm to incorporate the velocity propagation (i.e., computation of $\dot{X}_a$) as well as the computation of $\ddot{X}'_a$. Note that, as the result of the computation of Step I and for $i = n+1$, $\dot{X}_a$ and X'_a are obtained as

$$\dot{X}_a = \begin{bmatrix} {}^1\omega(n+1) \\ {}^1V(n+1) \end{bmatrix} \quad \text{and} \quad \ddot{X}'_a = \begin{bmatrix} {}^1\dot{\omega}'(n+1) \\ {}^1\dot{V}'(n+1) \end{bmatrix} \tag{31}$$

The cost of serial implementation of RAC from Eqs. (1)-(4) and (16)-(13) is obtained as $(180m + 152a)n - (27m + 21a)$. This includes the cost of serial computation of forward kinematics and Jacobian but the cost of linear system solution in Eq. (14) is not included.

The parallel computation of the algorithm in Table III requires the solution of a set of FOLRs in both directions, i.e., with increasing and decreasing indices as in Eqs. (5) and (7). The detailed computation of these FOLRs is not discussed and they are referred to be performed in a similar way as either in Eq. (5) or Eq. (7). The parallel computation of the algorithm is similar to the one in [32] for parallel computation of the N-E algorithm and is performed as follows:

Step I:

a. For i = 1 to n, Do in Parallel

$$PR_i : a(i+1) = {}^1Z(i)\dot{Q}(i) \tag{32}$$

End-Do Parallel

b. Parallel compute as in Eq. (5) the FOLR:

$${}^1\omega(i+1) = {}^1\omega(i) + a(i+1) \tag{33}$$

c. For $i = 1$ to n, Do in Parallel

$$PR_i : \text{ Send } {}^1\omega(i+1) \text{ to } PR_{i+1} \tag{34}$$

End-Do Parallel

d. For $i = 1$ to n, Do in Parallel

$$PR_i : b(i+1) = {}^1\omega(i+1) \times {}^1P(i) \tag{35}$$

End-Do Parallel

e. Parallel compute as in Eq. (5) the FOLR:

$${}^1V(i+1) = {}^1V(i) + b(i+1) \tag{36}$$

f. For $i = 1$ to n, Do in Parallel

$$PR_i : c(i+1) = {}^1\omega(i) \times a(i+1) \tag{37}$$

End-Do Parallel

g. Parallel compute as in Eq. (5) the FOLR:

$${}^1\dot{\omega}'(i+1) = {}^1\dot{\omega}'(i) + c(i+1) \tag{38}$$

h. For $i = 1$ to n, Do in Parallel

$$PR_i : d(i+1) = {}^1\dot{\omega}'(i+1) \times {}^1P(i) + {}^1\omega(i+1) \times b(i+1) \tag{40}$$

End-Do Parallel

e. Parallel compute as in Eq. (5) the FOLR:

$${}^1\dot{V}'(i+1) = {}^1\dot{V}'(i) + d(i+1) \tag{41}$$

The communication in Eq. (35) can be overlapped with the computation of Eqs. (36)-(37). Also, in Eqs. (35) and (37) the VC operations are performed by using 6 PE's but in Eq. (40) it is more efficient to perform the two VC operations in parallel by using 3 PE's for each VC operation. The computation cost of this step, by using 3 PE's for computation of SV and VA operations is obtained as $\lceil log_2 n \rceil (4a) + (5m + 4a)$.

As a result of the computation of forward kinematics and Jacobian, and step I, in addition to angular velocity propagation, the vectors $X_a, \dot{X}_a$, and $\ddot{X}'_a$ are computed and reside in the memory of the nth SIMD processor. This data can then be sent to the IH or any other processor which handles the computation of $\ddot{X}_c$ from Eq. (11) as well as the solution of the linear system in Eq. (14) for $\ddot{Q}$. Note that, the computation in Eq. (11) is not intensive and, also, as stated before, the solution of Eq. (14), for 6-DOF spherical wrist arms, can be obtained very efficiently.

The computation of Step II by SIMD processors is triggered by receiving the vector $\ddot{Q}$ and is performed as follows.

Step II:

a. For $i = 1$ to n, Do in Parallel

$$PR_i : e(i+1) = {}^1Z(i)Q(i) + {}^1\dot{\omega}'(i+1) \tag{42}$$

End-Do Parallel

b. Parallel compute as in Eq. (5) the FOLR:

$${}^1\dot{\omega}(i+1) = {}^1\dot{\omega}(i) + e(i+1) \tag{43}$$

c. For $i = 1$ to n, Do in Parallel

$$PR_i : g(i+1) = {}^1\dot{\omega}(i+1) \times {}^1P(i) + {}^1\dot{V}'(i+1) \tag{44}$$

End-Do Parallel

d. Parallel compute as in Eq. (5) the FOLR:

$${}^1\dot{V}(i+1) = {}^1\dot{V}(i) + g(i+1) \tag{45}$$

e. For $i = 1$ to n, Do in Parallel

$$PR_i : \text{ Send } \dot{V}(i+1) \text{ to } PR_{i+1} \tag{46}$$

$$PR_i : {}^1S(i) = O(1, i+1)\, {}^{i+1}S(i) \tag{47}$$

$$PR_i : {}^1\dot{V}_c(i) = {}^1\dot{V}(i) + {}^1\dot{\omega}(i+1) \times {}^1S(i) + {}^1\omega(i+1) \times ({}^1\omega(i+1) \times {}^1S(i)) \tag{48}$$

End-Do Parallel

The cost of this step, by maximizing the PE's utilization, is obtained as $(2a)\lceil log_2 n\rceil + (9m + 8a)$. The communication in Eq. (46) can be overlapped with the computation of Eq. (47) and part of the computation of Eq. (48).

Step III:

a. For $i = 1$ to n, Do in Parallel

$$PR_i : {}^{i+1}\omega(i+1) = O(i+1,1)\,{}^1\omega(i+1) \text{ and } {}^{i+1}\dot{\omega}(i+1) = O(i+1,1)\,{}^1\dot{\omega}(i+1) \tag{49}$$

$$PR_i : {}^{i+1}N(i) = {}^{i+1}I(i)\,{}^{i+1}\dot{\omega}(i+1) + {}^{i+1}\omega(i+1) \times ({}^{i+1}I(i)\,{}^{i+1}\omega(i+1)) \tag{50}$$

$$PR_i : {}^1F(i) = m(i)\,{}^1\dot{V}_c(i) \text{ and } {}^1N(i) = O(1,i+1)\,{}^{1i+1}N(i) \tag{51}$$

End-Do Parallel

b. Parallel compute as in Eq. (7) the FOLR:

$${}^1f(i) = {}^1F(i) + {}^1f(i+1) \tag{52}$$

c. For $i = 1$ to n, Do in Parallel

$$PR_i : h(i) = {}^1N(i) + {}^1S(i) \times {}^1F(i) + {}^1P(i) \times {}^1f(i+1) \tag{53}$$

End-Do Parallel

d. Parallel compute as in Eq. (7) the FOLR:

$$^{1}n(i) = h(i) + \ ^{1}n(i+1) \tag{54}$$

e. For $i = 1$ to n, Do in Parallel

$$PR_i : \tau(i) = \ ^{1}Z(i).\ ^{1}n(i) \tag{54}$$

End-Do Parallel

The two MV operations in Eq. (49) are performed in parallel and the SV and MV operations in Eq. (51) are overlapped. The computation cost of this step, by maximizing the PE's utilization in Eqs. (50) and (53) is obtained as $(2a)\lceil log_2 n\rceil + (15m + 14a)$.

Adding the cost of the above three steps plus the cost of parallel computation of forward kinematics and Jacobian, the total computational cost of parallel implementation of RAC is obtained as $(6m + 13a)\lceil log_2 n\rceil + (33m + 29a)$. For $n = 6$ and $n = 8$, the outlined parallel implementation of RAC represents a speedup of 16.3 and 21.9 over the serial implementation as described in Eqs. (1)-(4) and (16)-(30). The significance of this speedup in the computation is more appreciated when noted that the serial algorithm for RAC is also highly efficient since it exploits the synergism in the different computations and avoids the explicit computation of $\dot{J}_E$.

If parallelism in matrix-vector operations is not exploited then the computation cost of parallel implementation of RAC, by exploiting parallelism at the problem level only, is obtained as $(27m + 45a)\lceil \log_2 n\rceil + (147m + 104a)$. For $n = 6$ and $n = 8$, this represents a speedup of about 4.16 and 5.58, respectively. Again, it can be seen that the multilevel parallel computation leads to a significant improvement in achievable speedup in the computation.

V. Discussion and Conclusion

In this chapter we presented the design of ASPARC and evaluated its performance for two computational problems. At the problem level, ASPARC represents a practical and optimal parallel architecture for implementation of the parallel $O(\log n)$ algorithms for the kinematic and dynamic problems. At lower computational level, the design is not optimized for any specific operation but rather is based on efficient tradeoffs to exploit maximum parallelism in various matrix-vector operations while reducing the complexity of the architecture and maintaining its flexibility and reconfigurability.

The two examples in Sec. IV clearly illustrate that a multilevel approach to parallel computation of this class of problems can lead to a significant speedup even for problems with small size. Due to the lack of space, we did not discuss the implementation of more advanced $O(\log n)$ parallel algorithms for forward dynamics and operational space mass matrix and its inverse. It should be mentioned, however, that for these algorithms an even greater speedup in matrix-vector operations can be achieved since they involve the operations on larger matrices and vectors. Also, they involve more computation intensive operations, e.g., linear system solution, which allow a better utilization of the SIMD parallel processors and thus a greater speedup in their computations.

Another important problem for the control of redundant arms is the Jacobian inversion, that is, the solution of Eq. (14). As stated before, this solution is usually both task and arm dependent. However, if any specific method for its solution is formulated in terms of numerical procedures then the computing power of ASPARC can be used to accelerate the solution. One particularly appropriate method for solution of Eq. (14) on ASPARC is based on the use of Singular Value Decomposition (SVD) of the Jacobian matrix. Though for many practical applications the SVD method can represent the optimal choice, it is not used for real-time control due to its prohibitive computation cost [38]. However, the computation of SVD for Jacobian can be efficiently parallelized and implemented on ASPARC. In this regard, a relevant result is the work in [37] wherein parallel computation of SVD on two-dimensional processor arrays is analyzed. Note that the Jacobian matrix is a $6 \times n$ matrix. The PCE can be configured to emulate a two-dimensional array of $6 \times n$ PE's with a nearest neighbor interconnection. The parallel algorithm for computation of SVD as well as the parallel algorithms for computation of forward dynamics and operational space mass matrix and its inverse will be presented in forthcoming papers.

Appendix: Computation of Time Derivative of Jacobian

Jacobian matrix can be presented as ${}^{C}J_R$ [15] where R denotes the velocity reference point and C the coordinate frame in which the elements of Jacobian matrix are described. For the control purpose, ${}^{1}J_E$ and ${}^{E}J_E$, and their time derivatives are of interest where E stand for EE frame, i.e., frame $n+1$. In the following, the coordinate-free form of Jacobian, that is, Jacobian vectorial matrix J_E is considered and its time derivative $\dot{J}_E$ is derived. This derivation is independent of the frame in which the elements of Jacobian matrix are described. The desired projection of $\dot{J}_E$, i.e., ${}^{1}\dot{J}_E$ or ${}^{E}J_E$, can be simply obtained by projecting the elements of $\dot{J}_E$ in frame 1 or $n+1$. Note, however, that the time derivative should be taken in inertia frame.

The Jacobian vectorial matrix J_E and its elements are given as

$$J_E = \begin{bmatrix} J_r \\ J_t \end{bmatrix} \quad \text{and} \quad \dot{J}_E = \begin{bmatrix} \dot{J}_r \\ \dot{J}_t \end{bmatrix} \tag{A1}$$

$$J_r = \begin{bmatrix} Z(1) & Z(2) & \cdots & Z(n) \end{bmatrix}$$

$$J_t = \begin{bmatrix} Z(1) \times P(n+1,1) & Z(2) \times P(n+1,2) \cdots & Z(n) \times P(n+1,n) \end{bmatrix} \tag{A2}$$

The ith element of matrices J_r and J_t are denoted as J_{ri} and J_{ti}. $\ddot{X}'_a$ is given as

$$\ddot{X}'_a = \begin{bmatrix} \dot{\omega}'(n+1) \\ \dot{V}'(n+1) \end{bmatrix} = \begin{bmatrix} \dot{J}_r \\ \dot{J}_t \end{bmatrix} \dot{Q}, \text{ or} \tag{A3}$$

$$\dot{\omega}'(n+1) = \dot{J}_r \dot{Q} \text{ and } \dot{V}'(n+1) = \dot{J}_t \dot{Q} \tag{A4}$$

In order to compute $\dot{J}_E$ the time derivatives of vectors $Z(i)$ and $P(n+1,i)$ need to be computed.

Given two frames A and B, the time derivatives of a vector V taken in frame A and in frame B are related as

$$\frac{d^A}{dt} V = \frac{d^B}{dt} V + \omega_B^A \times V \tag{A5}$$

where superscript indicates the frame in which the time derivative is taken and ω_B^A stands for the angular velocity of frame B with respect to frame A. In the following the absence of superscript denotes that the time derivative is taken in the inertia frame and that the angular velocity is also described with respect to the inertia frame.

From Eq. (A5), the time derivative of the vector J_{ri} can be computed as

$$\frac{d}{dt}J_{ri} = \dot{J}_{ri} = \frac{d}{dt}Z(i) = \dot{Z}(i) = \frac{d^i}{dt}Z(i) + \omega(i) \times Z(i) \tag{A6}$$

Since the vector $Z(i)$ is constant in frame i, Eq. (A6) simplifies as

$$\dot{J}_{ri} = \dot{Z}(i) = \omega(i) \times Z(i) \tag{A7}$$

From Eqs. (A4) and (A7), it follows that

$$\dot{\omega}'(n+1) = \dot{J}_r\dot{Q} = \sum_{i=1}^{n} \dot{J}_{ri}\dot{Q}(i) = \sum_{i=1}^{n} \dot{Z}(i)\dot{Q}(i) = \sum_{i=1}^{n} \omega(i) \times Z(i)\dot{Q}(i) \tag{A8}$$

The last summation in Eq. (A8) can be computed recursively as

For $i = 1$ to n, Do

$$\dot{\omega}'(i+1) = \dot{\omega}'(i) + \omega(i) \times Z(i)\dot{Q}(i) \tag{A9}$$

End-Do

which is exactly the recursion given in Eq. (18) of Table II.

To compute the time derivative of J_{ti} the vector $P(n+1,i)$ is first written as

$$P(n+1,i) = \sum_{j=i}^{n} P(j+1,j) \tag{A10}$$

The time derivative of vector $P(i+1,i)$ is computed as

$$\frac{d}{dt}P(i+1,i) = \dot{P}(i+1,i) = \frac{d^{i+1}}{dt}P(i+1,i) + \omega(i+1) \times P(i+1,i) \qquad (A11)$$

Since the vector $P(i+1,i)$ is constant in frame $i+1$, Eq. (A11) simplifies as

$$\dot{P}(i+1,i) = \omega(i+1) \times P(i+1,i) \qquad (A12)$$

from which the time derivative of vectors $P(n+1,i)$ and J_{ti} are computed as

$$\dot{P}(n+1,i) = \sum_{j=i}^{n} \dot{P}(j+1,j) = \sum_{j=i}^{n} \omega(j+1) \times P(j+1,j) \qquad (A13)$$

$$\begin{aligned}\dot{J}_{ti} &= Z(i) \times \dot{P}(n+1,i) + \dot{Z}(i) \times P(n+1,i) \\ &= Z(i) \times \big(\sum_{j=i}^{n} \omega(j+1) \times P(j+1,j)\big) + (\omega(i) \times \ Z(i)) \times P(n+1,i)\end{aligned} \qquad (A14)$$

From Eq. (A4) and (A14), it follows that

$$\begin{aligned}\dot{V}'(n+1) = \dot{J}_t\dot{Q} &= \sum_{i=1}^{n} \dot{J}_{ti}\dot{Q}(i) \\ &= \sum_{i=1}^{n}\big(\sum_{j=i}^{n} Z(i)\dot{Q}(i) \times (\omega(j+1) \times P(j+1,j)\big) \\ &+ \sum_{j=i}^{n}(\omega(i) \times Z(i)\dot{Q}(i)) \times P(j+1,j))\end{aligned} \qquad (A15)$$

Using the identity $\sum_{i=1}^{n}\sum_{j=i}^{n} = \sum_{j=1}^{n}\sum_{i=1}^{j}$, Eq. (A15) is written as

$$
\begin{aligned}
\dot{V}'(n+1) &= \sum_{j=1}^{n}\sum_{i=1}^{j} Z(i)\dot{Q}(i) \times (\omega(j+1) \times P(j+1,j)) \\
&+ (\omega(i) \times Z(i)\dot{Q}(i)) \times (P(j+1,j)) \\
&= \sum_{j=1}^{n}(\sum_{i=1}^{j} Z(i)\dot{Q}(i)) \times (\omega(j+1) \times P(j+1,j)) \\
&+ \sum_{j=1}^{n}(\sum_{i=1}^{j} \omega(i) \times Z(i)\dot{Q}(i)) \times P(j+1,j)
\end{aligned} \tag{A16}
$$

From Eqs. (16) and (A8)-(A9), it follows that

$$
\dot{V}'(n+1) = \sum_{j=1}^{n} \omega(j+1) \times (\omega(j+1) \times P(j+1,j)) + \dot{\omega}'(j+1) \times P(j+1,j) \tag{A17}
$$

The summation in Eq. (A17) can be computed recursively as

For $j = 1$ to n, Do

$$
\dot{V}'(j+1) = \dot{V}'(j) + \omega(j+1) \times (\omega(j+1) \times P(j+1,j)) + \dot{\omega}'(j+1) \times P(j+1,j) \tag{A18}
$$

End-Do

which is exactly the recursion given in Eq. (19) of Table II.

Note that, $\dot{J}_E$ can be explicitly computed from Eqs. (A7) and (A14). This computation, however, is inefficient since in addition to computation in Eqs. (A7) and (A14) the matrix-vector operation in Eq. (A3) needs to be performed. In particular, while Eq. (A7) can be directly computed in $O(n)$ steps, the computation of Eq. (A14) in $O(n)$ steps requires the summation to be performed in a reverse order. This further illustrates the greater efficiency of Eqs. (18)-(19) wherein the synergism in computation of $\dot{\omega}'(i+1)$ and $\dot{V}'(i+1)$, and in computation of $\dot{\omega}(i+1)$ and $\dot{V}(i+1)$ is also exploited.

Similar to other kinematic and dynamic problems, the time derivative of Jacobian can also be computed in $O(\log n) + O(1)$ steps on the same SIMD parallel architecture with n processors interconnected through SENN structure. To see this, let us consider the computation of ${}^1\dot{J}_E$. Also, let us assume that the vectors ${}^1Z(i)$, ${}^1P(n+1,i)$, and ${}^1\omega(i+1)$, for $i = 1$ to n, are already computed in parallel in $O(\log n) + O(1)$ steps as described in Section IV. With n processors and from Eq. (A7), the terms ${}^1\dot{J}_{ri}$, for $i = 1$ to n, can be computed in one step. For computation of ${}^1\dot{J}_{ti}$ from Eq. (A14), first the terms ${}^1\omega(j+1) \times {}^1P(n+1,j)$, for $i = 1$ to n, are computed in parallel and in one step. The summations $\sum_{j=i}^{n} {}^1\omega(j+1) \times {}^1P(n+1,j)$, for $j = n$ to 1, can then be computed in $\lceil \log_2 n \rceil$ steps. Finally the terms ${}^1\dot{J}_{ti}$, for $i = 1$ to n, can be computed in one step, leading to a total computational complexity of $O(\log n) + O(1)$. Note, also, that with n processors the matrix-vector multiplication in Eq. (A3) can be performed in $O(\log n) + O(1)$ steps. However, as can be seen, for parallel computation it is also more efficient and straightforward to simply compute Eq. (18)-(19) in parallel.

Acknowledgment

The research described in this paper was carried out by the Jet Propulsion Laboratory, California Institute of Technology, under contract with the National Aeronautics and Space Administration (NASA). The ASPARC is the subject of a pending U.S. patent owned by NASA.

References

1. J. Angeles,"On Numerical Solution of Inverse Kinematic Problem," Int. J. Robotics Res. Vol. 4(2), pp. 21-37, 1985.

2. W.W. Armstrong,"Recursive Solution to the Equation of Motion of an N-Link Manipulator," Proc. 5th World Congress on Theory of Machines and Mechanisms, pp. 1343-1346, 1979.

3. J. Barhen,"Hypercube Ensembles: An Architecture for Intelligent Robots," in *Computer Architectures for Robotics and Automation*, Ed. J.H. Graham, Gordon and Breach Science Publishers, 1987.

4. A.K. Bejczy, T.J. Tarn, and X. Yun,"Robust Robot Arm Control with Nonlinear Feedback," Proc. IFAC Sym., Barcelona, 1985.

5. S. Butner, Y. Wang, A. Mangaser, and S. Jordan,"Design and Simulation of RIPS: An Advanced Robot Control System," Proc. IEEE Int. Conf. Robotics and Auto., pp. 470-474, April 1988, Philadelphia, PA.

6. R. Featherstone,"The Calculation of Robot Dynamics Using Articulated-Body Inertia," Int. J. Robotics Research, Vol. 2(1), pp. 13-30, 1983.

7. A. Fijany,"A Parallel O(log n) Algorithm for Computation of Manipulator Forward Dynamics," submitted to IEEE Trans. Robotics and Automation. Also, in JPL Eng. Memo. EM. 1258-92-343, April 1992.

8. A. Fijany,"Parallel O(log n) Algorithms for Dynamic Simulation of Closed-Chain Multiple Robot Arms," In preparation.

9. A. Fijany,"New O(n) Algorithms for Computation of Operational Space Mass Matrix and its Inverse," In preparation.

10. A. Fijany, *Parallel Algorithms and Architectures for Robotics.* Ph.D. Diss., University of Paris XI (Orsay), Sept. 1988.

11. A. Fijany and A.K. Bejczy,"Parallel Computation of Manipulator Inverse Dynamics," J. Robotic Systems, Vol. 8(5), pp. 599-635, Oct. 1991.

12. A. Fijany and A.K. Bejczy,"Techniques for Parallel Computation of Mechanical Manipulator Dynamics. Part II: Forward Dynamics," in *Advances in Control and Dynamic Systems, Vol. 40: Advances in Robotic Systems Dynamics and Control*, C.T. Leondes (Ed.), pp. 357-410, Academic Press, March 1991.

13. A. Fijany and A.K. Bejczy,"Parallel Algorithms and Architecture for Computation of Robot Manipulator Forward Dynamics," Proc. IEEE Int. Conf. Robotics and Auto., pp. 1156-1162, April 1991, Sacramento, CA.

14. A. Fijany and A.K. Bejczy,"A Class of Parallel Algorithms for Computation of Manipulator Inertia Matrix," IEEE Trans. Robotics Auto., Vol. 5(5), pp. 600-615, Oct. 1989.

15. A. Fijany and A.K. Bejczy,"Efficient Jacobian Inversion for the Control of Simple Robot Manipulators," Proc. IEEE Int. Conf. Robotics and Automation, pp. 999-1007, April 1988, Philadelphia, PA.

16. A. Fijany and J.G. Pontnau,"Parallel Computation of the Jacobian for Robot Manipulators," Proc. IASTED Int. Conf. Robotics and Auto., 1987.

17. A. Fijany and R.E. Scheid,"Fast Serial and Parallel Conjugate Gradient Algorithms for Computation of the Manipulator Forward Dynamics," submitted to J. of Robotic Systems. Also, in JPL Eng. Memo, EM 343-1196, Aug. 1991.

18. E. Freund, "Fast Nonlinear Control with Arbitrary Pole Placement for Industrial Robots and Manipulators," Int. J. Robotics Res., Vol. 1(1), 1982.

19. A. Gibbons and W. Rytter, *Efficient Parallel Algorithms.* Cambridge Univ. Press, 1988.

20. R.W. Hockney and C.R. Jesshope, *Parallel Computers.* Adam Hilger, 1981.

21. K. Hwang and F.A. Briggs, *Computer Architecture and Parallel Processing.* McGraw-Hill, 1984.

22. L.H. Jamieson,"Characterizing Parallel Algorithms," in *The Characteristics of Parallel Algorithms*, L.H. Jamieson et al (Eds.), The MIT Press, 1987.

23. H. Kasahara, H. Fujii, and M. Iwata,"Parallel Processing of Robot Motion Simulation," Proc. 10th IFAC World Congress, July 1987.

24. H. Kasahara and S. Narita,"Parallel Processing of Robot-Arm Control Computation on a Multimicroprocessor," IEEE J. Robotics and Automation, Vol. RA-1(2), June 1985.

25. O. Khatib, *Commande Dynamique dans l'Espace Operationnel des Robots Manipulateurs en Presence d'Obstacles.* Ph.D diss., ENSAE, 1980.

26. P.M. Kogge,"Parallel Solution of Recurrence Problems," IBM J. Res. Develop.,

Vol. 18, March 1974.

27. L. Kronsjo, *Computational Complexity of Sequential and Parallel Algorithms*, John Wiley and Sons, 1985.

28. Kung, H.T.,"Why Systolic Architectures?," IEEE Computer, Jan. 1982.

29. Kung, S.Y.,"On Supercomputing with Systolic/Wavefront Array Processors," Proc. of The IEEE, Vol. 72(7), 1984.

30. L.H. Lathrop,"Parallelism in Manipulator Dynamics," Int. J. Robotics Res., Vol.4(2), 1985.

31. C.S.G. Lee and P.R. Chang,"Efficient Parallel Algorithms for Robot Forward Dynamics Computation," IEEE Trans. Syst., Man, and Cybern., Vol. 18(2), pp. 238-251, March/April 1988.

32. C.S.G. Lee and P.R. Chang,"Efficient Parallel Algorithms for Robot Inverse Dynamics Computation," IEEE Trans. Syst., Man, and Cybern., Vol. 16(4), pp. 532-542, July/August 1986.

33. Y.L.C. Ling, P. Sadayappan, K.W. Olson, D.E. Orin,"A VLSI Robotics Vector Processor for Real-Time Control," Proc. IEEE Int. Conf. Robotics and Auto., pp. 303-308, April 1988, Philadelphia, PA.

34. J.Y.S. Luh, M.W. Walker, and R.P.C. Paul,"Resolved Acceleration Control of Mechanical Manipulators," IEEE Trans. Auto. Control, Vol. AC-25(3), 1980.

35. J.Y.S. Luh, M.W. Walker, and R.P.C. Paul,"On-Line Computational Scheme for Mechanical Manipulator," ASME J. Dynamic Syst., Meas., Control, Vol. 102, pp. 69-76, June 1980.

36. J.Y.S. Luh and C.S. Lin,"Scheduling of Parallel Computation for a Computer Controlled Mechanical Manipulators," IEEE Trans. Syst., Man, and Cybern., Vol. SMC-12(2), March 1982.

37. F. T. Luk,"A Rotation Method for Computing the QR-Decomposition," SIAM J. Sci. Stat. Comput., Vol. 7(2), pp. 452-459, 1896.

38. A.A. Maciejewski and C.A. Klein,"The Singular Value Decomposition: Computation and Application to Robotics," Int. J. Robotics Res., Vol. 8(6), pp. 63-79, Dec. 1989.

39. S. McMillan, P. Sadayappan, and D.E. Orin,"Efficient Dynamic Simulation of

Multiple Manipulator Systems with Singular Configuration," Proc. IEEE Int. Conf. Robotics and Auto., May 1992.

40. J. Miklosko and V.E. Kotov (Eds.), *Algorithms, Software and Hardware of Parallel Computers*, Springer-Verlag, 1984.

41. D.E. Orin et al,"Pipeline/Parallel Algorithms for the Jacobian and Inverse Dynamics Computation," IEEE Int. Conf. Robotics and Automation, March 1985.

42. D.E. Orin, K.W. Olson, H-H. Chao,"Systolic Architectures for Computation of the Jacobian for Robot Manipulators," in *Computer Architectures for Robotics and Automation*, Ed. J.H. Graham, Gordon and Breach Science Pub., 1987.

43. R.P.C. Paul, "Modeling, Trajectory Calculation, and Servoing of a Computer Controlled Arm," Stanford Univ., AI Memo 177, Sept. 1972.

44. G. Rodriguez,"Kalman Filtering, Smoothing and Recursive Robot Arm Forward and Inverse Dynamics," IEEE J. Robotics and Automation, Vol. 3(6), pp. 624-639, Dec. 1987.

45. G. Rodriguez and K. Kreutz-Delgado,"Spatial Operator Factorization and Inversion of Manipulator Mass Matrix", IEEE Trans. Robotics and Automation, Vol. 8 (1), pp. 65-76, Feb. 1992.

46. G. Rodriguez, K. Kreutz, and A. Jain,"A Spatial Operator Algebra for Manipulator Modeling and Control," Int. J. Robotics Research, vol. 10(4), pp. 371-381.

47. L. Snyder et al (Eds.), *Algorithmically Specialized Parallel Computers*. Academic Press, 1985.

48. H. Stone,"Parallel Processing with the Perfect Shuffle," IEEE Trans. on Computer, Vol. 20(2), pp. 153-161.

49. M.W. Walker and D.E. Orin, "Efficient Dynamic Computer Simulation Of Robotic Mechanism," ASME J. Dynamic Systems, Measurement, and Control, Vol. 104, pp. 205-211, 1982.

50. Y. Wang et al, The 3DP: A Processor Architecture for Three-Dimensional Applications," IEEE Computer, Vol. 25(1), pp. 25-36, Jan. 1992.

51. Y-F Zheng and H. Hemami,"Computation of Multibody System Dynamics by a Multiprocessor Scheme," IEEE Trans. Syst., Man, Cybern., Vol. SMC-16(1),

Jan. 1986.

52. *32-bit Microprogrammable Products Data Book*, Advanced Micro Devices, 1988.

53. *32-bit Floating-Point DSP Processors*, EDN, pp. 126-146, Feb. 1991.

CHAPTER 5

MULTIPROCESSOR BASED TRAJECTORY GENERATION FOR MULTIPLE ROBOTIC DEVICES

Rick Guptill[+] and Shaheen Ahmad[++]
[+]Silma, Inc.
Cupertino, CA

[++]Real-time Robot Control Laboratory
School of Electrical Engineering
Purdue University
West Lafayette, IN 47907

Abstract

In this chapter we review a trajectory computation scheme for single and multiple robotic devices using homogeneous transforms. A concise method of specifying and generating the coordinating motion for multiple robotic devices using homogeneous transforms is given.

The components of a multiprocessor robot controller capable of communication and kinematic motion coordination with multiple robotic devices is also addressed in this chapter. We first discuss the class of multiprocessor system we will target and the expected configuration and function of each processor. We then present a conceptually new approach as to how to accomplish real-time task scheduling as a function of the changing Cartesian relationships in the robotic task. By introducing the concept of a Transform Task List, we allow our motion system to run efficiently on a controller with one or many processors. Finally we simulate the multiprocessor system and give examples of speed-up using our design.

1.0 Introduction

Currently in industry there are a number of applications which can be regarded as multi-robot coordination problems. These include a multi-motion device welding system which can consist of a robot, a motion track and a welding table positioner. Here the welding table continuously orients the part so as to make the weld surface normal parallel to the vertical. The robot manipulates the weld gun to follow the desired weld trajectory (Ahmad & Luo 89) see Figure 1. The track is moved to keep the robot in an optimal orientation and reach configuration. Another multi-robot application includes multi-robot spot welding of car frames which are in motion. Here the coordination of multiple robots must be such that the moving cars are spot welded while avoiding collisions between the car fixtures and other robots. Yet another multi-robot application, which we will use as an example later in the chapter, is a multi-arm inspection system of large composite panels using ultrasound, see Figure 2.

Despite tremendous work in the area of multi-robot feedback control schemes (Ahmad and Guo 89), (Hyati 86), the programming environment to implement multi-robot tasks are as yet uncommon. Although a few experimental systems in university environments have been described in the literature. Two such multi-robot package are the NASA - McGill's KALI programming system (Hayward and Hyati 88), and the Cybotech-Purdue's multi-arm programming system (Guptill and Stahura 87).

In order to understand the computational requirements of single and multiple robot trajectory systems it is necessary to understand the mechanics of robot motion specification and execution. The single and multiple robot motion system can be understood in terms of the motion control primitives used to specify target positions of the robot end effector with respect to some reference frame. Usually the control primitives are generated after collision free paths have been determined in an off-line manner (see, Guptill and Ahmad 90). Therefore collision avoidance, singularity avoidance and motion limit checking are not usually performed by the on line trajectory generator. These issues will not be addressed in this chapter.

There has been a tremendous amount of work done in the area of single robot coordination (Ahmad and Besant 84, Paul 81, Hayward 83, Hayward and Paul 84), using 4×4 homogeneous transformation matrices. In this chapter we show how 4×4 transform matrices together with motion coordination concepts schemes can be used to coordinate multiple robots.

The mechanization of the motion control primitives can be achieved by using the homogenous 4×4 matrix transforms to represent the relative geometric relationships between end-effectors and reference frames in the task space. This also allows for easy interface to sensors and subsequent modification of end effector motions based on sensory or programmed information. This idea has been exploited by a number of authors (see Ahmad and Besant (Ahmad and Besant 84), Paul (Paul 81), Hayward (Hayward 83), Hayward (Hayward and Paul 84)).

In order to clearly describe the mechanism of motion generation we will assume 'C' language constructs such as in data structures, pointers, functions and function call procedures.

The first part of this chapter describes robot motion specification and its representation in terms of data structures and transforms, most of this work can be found in Paul (Paul 81a) and also in (Hayward 83). The second part of this chapter describes a scheduling algorithm for multirobot trajectory execution in a multiprocessor environment.

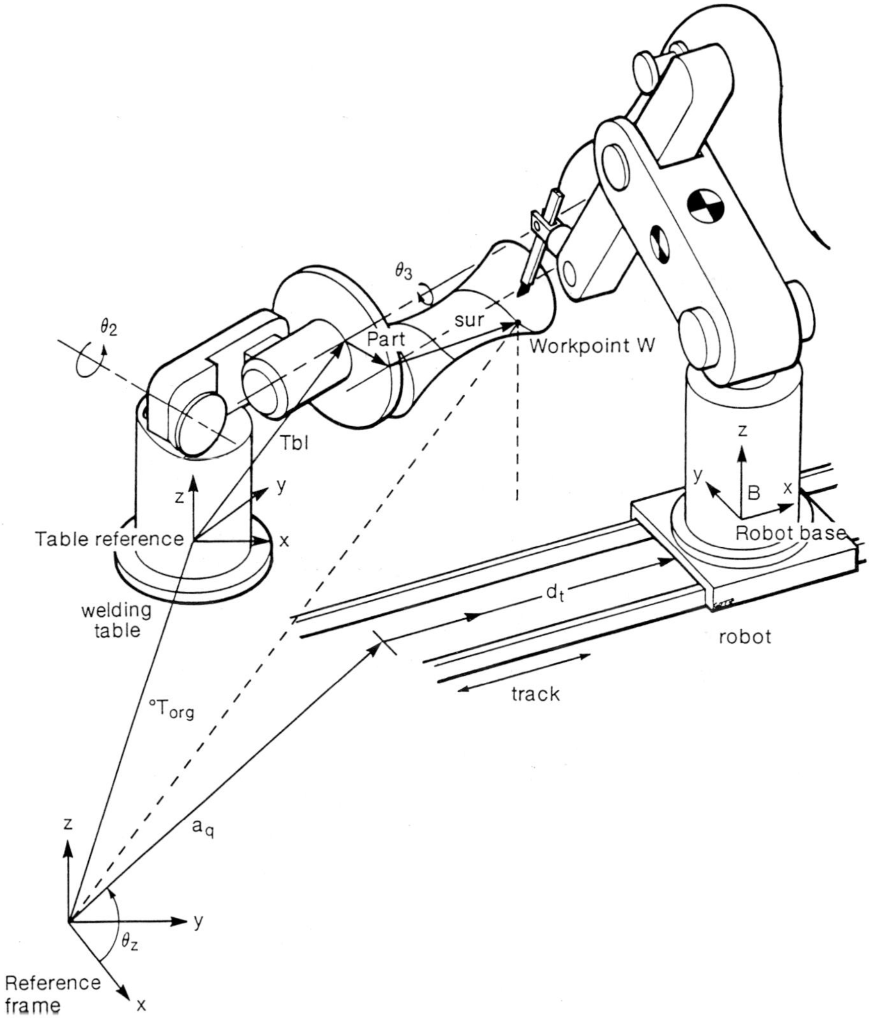

Figure 1. Multi-Robot Welding System

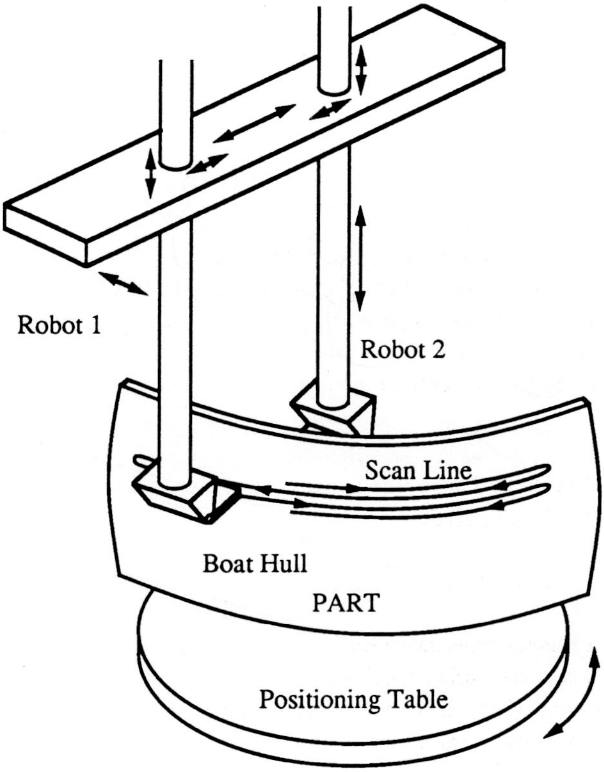

Figure 2. Multi-Robot Inspection System

2.0 Preliminaries: Data Structures and Notation

2.1a Data Structure

In this chapter, we will follow the C language convention to represent data structures, further we will take advantage of typesetting features to aid in the understanding of the variables used. For example, if we wish to define a data structure for a three dimensional vector, we write:

```
typedef struct vect {
        REAL x, y, z;
} VECTOR, *VECTOR_PTR;
```

where REAL is the data type of a floating point number, VECTOR defines the structure, and VECTOR_PTR defines a pointer to the VECTOR structure. A pointer to a structure contains the address of the structure in the computer memory. with the memory allocation for the structure is typically assigned by a dynamic memory allocator. To access a element of a structure given a pointer to the structure, the $\rightarrow$ operator is used.

2.1b Notation

A list of mathematical symbols which are used in this chapter follows. Their meaning will become clear in the following sections.

Symbol	Description
$D(r)$	The primary drive transform.
$D_s(p)$	The secondary drive transform.
Δ ()	Function which calculates the primary drive parameters.
Δ_s ()	Function which calculates the secondary drive parameters.
Γ ()	Function which solves for T_n for a given position equation.
h	A variable which changes from 0 to 1 during each movement.
K ()	Forward kinematic function of a robotic device.
K^{-1} ()	Inverse kinematic function of a robotic device.
Λ ()	Function which computes the matching transition coefficients.
L_T	The left side of the Cartesian transition equation.
M_B	Motion structure for position P_B.
M_C	Motion structure for position P_C.
p	A variable which changes from 0 to 1 each transition.
P_B	Position equation describing the current point B.
P_C	Position equation describing the current point C.
P_{new}	The new position read off the motion queue.

P_T	The Cartesian transition equation.
R_T	The right side of the Cartesian transition equation.
r	1 - h, a variable which changes from 1 to 0 during a movement.
T	The general symbol for homogeneous transform.
T_n	Homogeneous transform describing the last link of a robotic device.
$T_A(-t_{acc}-\tau)$	T_n at the sample period before the start of a motion transition.
$T_A(-t_{acc})$	T_n at the start of the transition.
$T_B(-t_{acc})$	T_n at the start of the transition of the P_B position.
$T_C(-t_{acc}-\tau)$	T_n at the sample period before the start of the transition of the P_C position.
$T_C(-t_{acc})$	T_n at the start of the transition of the P_C position.
T_{full}	T_n as returned by Γ () before K (). Its contents might contain more information than the K^{-1} () function can use.
$\vec{\theta}$	The general symbol for a joint variable vector.
$\vec{\theta}_A(-t_{acc})$	The current joint variables at the start of the transition.
$\vec{\theta}'_A(-t_{acc})$	The current joint velocities at the start of the transition.
$\vec{\theta}_B(-t_{acc})$	The joint variables at the start of the transition of P_B.
$\vec{\theta}'_B(-t_{acc})$	The joint velocities at the start of the transition of P_B.
$\vec{\theta}_C(-t_{acc})$	The joint variables at the start of the transition of P_C.
$\vec{\theta}'_C(-t_{acc})$	The joint velocities at the start of the transition of P_C.
$\vec{\theta}_{BC}$	$\vec{\theta}_B$ - $\vec{\theta}_C$
$\vec{\theta}'$	The joint velocities for the given device.
$\vec{\theta}_{last}$	The last sample period's $\vec{\theta}_{out}$.
$\vec{\theta}_{out}$	A joint variable vector equal in size to the number of axes of the robotic device containing the current joint variables to be sent to the robotic device.
t_{acc}	The acceleration / deacceleration time of the robotic device.
t_{seg}	The total time used to move to the next position.
τ	The trajectory sample time.
χ_A	The drive parameters calculated at the start of the transition.
$\chi'_B(-t_{acc})$	The drive velocities of the P_B position at the start of the transition.
$\chi'_C(-t_{acc})$	The drive velocities of the P_C position at the start of the transition.
χ_P	The primary drive parameters.
χ_T	The secondary Cartesian transition drive parameters.
χ_{temp}	Temporary drive parameters.

2.2 Homogeneous Transforms

To spécify a robot task, we first need a method of representing locations of objects and path points in the workspace. Given their ease of use and efficiency of representation, we will use homogeneous transforms to perform this function, as they can be used to describe the position and orientation between two coordinate frames (Paul 81a). A homogeneous transform consists of four three element vectors, **n**, **o**, **a**, and **p**. The orientation is described by the three direction cosines vectors and is stored in **n**, **o**, and **a**. The translational distance between the two defining frames is stored in the **p** vector, where the x, y, and z values are the location of the 'to' coordinate frame as measured in the 'from' frame reference.

We will assume homogeneous transforms are *functionally* defined. This means the transform value can be changed with a call to a C language function. The function can describe a mathematical path the robot is to follow, receive input from a sensor, or read a pre-defined stored path. The benefit of functionally defined transforms is the capability to modify the robot's path in the coordinate frame which is most convenient. The data structure we will use for a homogeneous transform is:

```
typedef struct homogeneous_transform {
        CHAR *name;                  /* user given name */
        REAL time_evaluated;
        VECTOR n,o,a,p;              /* actual data */
        struct trans *inverse;       /* inverse transform */
        FUNCTION_PTR defining_function;
} TRANSFORM, *TRANSFORM_PTR;
```

where FUNCTION_PTR is defined as a pointer to a C function returning an integer.

```
typedef int (*FUNCTION_PTR)();
```

The subroutine which dynamically allocates memory space for transform data structures also fills in the transform contents, allocates memory space for the name, and sets the *defining_function* to point to the defining subroutine if one exists. If the transform does not change over the course of the task, a pointer to the dummy C subroutine constant () is used for *defining_function*. Create_transform returns a pointer to the newly created transform. For example, to create a transform describing the base of a robot, we call

robot_base = create_transform(ROBOT BASE, Cartesian location, C function pointer); (1)

where the Cartesian location could be roll, pitch, and yaw angles, Euler angles, and/or positional information. This call also creates the inverse transform.

2.3 Motion Devices

Just as we used the TRANSFORM structure to represent the position and orientation difference between two coordinate frames, we will find it useful to abstract robotic motion devices and the functions and data structures associated with them into a single DEVICE data structure.

Before we define the DEVICE structure, however, let us first describe some variables and functions we will need to perform a robot task. First, each robotic device consists of one or more controllable axes of motion, or joints. We will store the current joint variables of each device in a JOINT structure. The number of axes of the device is included in the structure.

```
typedef struct joint_variables_info {
        REAL joint[MAX_JOINTS];
        INT number_of_axes;
} JOINT, *JOINT_PTR;
```

2.4 Forward and Inverse Kinematics

Homogeneous transforms can be used to describe the robotic motion devices we wish to coordinate. A robotic or motion device consists of a set of links connected by a set of joint actuators, these being either prismatic or revolute. By assigning a coordinate frame to the end of each link (Paul 81), we can define a set of homogeneous transforms, conventionally called $\mathbf{A}_i$ for the ith link, which define the relationship between the ith and $(i-1)$ links. The defining relationship consist of three parameters and the joint variable, depending on the type of joint. For a revolute joint θ is the variable and for a prismatic joint d is the controlled variable (Paul 81). These may be expressed as a concatenation of four homogeneous transformations.

$$A_i = \mathrm{Rot}(z, \theta)\ \mathrm{Trans}(a, 0, 0)\ \mathrm{Trans}(0, 0, d)\ \mathrm{Rot}(x, \alpha) \tag{2}$$

The link length a_i and the twist angle α_i are parameters of the link geometry (Paul 81a). In equation (2) Z is the axis around which joint is rotated by angle θ, similarly α_i is rotated about the x axis. The product of the above homogeneous transforms results in the general form of the $\mathbf{A}_i$ matrix

$$\mathbf{A}_i = \begin{bmatrix} \cos\theta & -\sin\theta\cos\alpha & \sin\theta\sin\alpha & a\cos\theta \\ \sin\theta & \cos\theta\cos\alpha & -\cos\theta\sin\alpha & a\sin\theta \\ 0 & \sin\alpha & \cos\alpha & d \\ 0 & 0 & 0 & 1 \end{bmatrix} \tag{3}$$

The position and orientation of the last link, T_n can be found by concatenating the n $\mathbf{A}_i$ homogeneous transforms. This is referred to as the *forward kinematic* transform. Given a set of joint variables we can uniquely determine T_n as,

$$T_n = A_1 \; A_2 \; \cdots \; A_n \tag{4}$$

We will use T_n to represent the Cartesian position and orientation of the last link of the motion device. This is to avoid the need for a symbol dependent on the device's degrees of freedom.

The *inverse kinematic* solution finds a set of joint variables corresponding to a given T_n. In general, this mapping is not one to one and a configuration variable is used to decide on one solution. Various methods exist to find the inverse kinematic solution for a given robot (Paul 81a) (McCarthy 86) (Gupta 86) (Paul 81b). Pieper (Pieper 68) first established that robots with the last three axes intersecting have a closed form inverse kinematic solution. Many industrial robots meet this criteria and Paul (Paul 81a) has established a systematic method for obtaining a closed form inverse kinematic solution for this class of robots. In controlling multiple robotic devices, we often encounter devices with less than six axes and have found the method in Paul (Paul 81a) to work well. If a closed form can not be found, numerical solutions exist which can be used with the motion control system described here (Goldenberg et. al. 85).

We will refer to the forward kinematic operation as K () and the inverse kinematic operation as K^{-1} (). In general the K^{-1} () operation to find the joint angles θ will need a configuration variable to chose one solution and a call to the function reflects that.

$$\vec{\theta} = K^{-1}(T_n, \text{configuration}) \tag{5}$$

the configuration variable is used to decide whether shoulder up or shoulder down or left or right arm configurations are to be employed. In this discussion we have assumed the robot is non-redundant.

2.5 Position Equations

By including the T_n transform of a device in a concatenation of transforms describing intermediate coordinate frames in a work space, we form a *position equation* (Paul 81a).

$$P_1 : [\text{base}][T_n][\text{gripper}] = [\text{table}][\text{block}] \tag{6}$$

TOOL : [gripper]

A position equation is used to specify the desired robotic device end-effector position and orientation in space. When the last link of the device is moved into the desired position and orientation as recorded by the T_n transform of the device, the position equation equality is satisfied. The transform specified as the TOOL, in this case the [gripper] transform, is the coordinate frame in which straight line movement will take place if it is specified.

Given that we wish to coordinate more than one robotic device, we will use the pointer to the device structure instead of the T_n transform pointer in the position equation, to indicate that we wish to control that device. If we use the device pointer name pointing to its T_n transform, we don't want to coordinate the device, but instead we just want to use its Cartesian position and orientation every sample time as if it were a constant transform. An example of this usage shows up in Section 4.1.

The above position equation for the PUMA robot is then

$$P_1 : [\text{base}][\text{puma}][\text{gripper}] = [\text{table}][\text{block}] \tag{7}$$

TOOL: [gripper]

given puma is a pointer to the PUMA device structure.

The internal representation of a position equation is a doubly linked list (Paul 83a), as shown in Figure 3.

A user routine takes as input transform pointers and device pointers and forms the doubly linked list. The data structures involved are a position header, to which the position pointer points to, and position link structures, which point to the transforms specified. The data structures for a position link and a position equation are as follows.

```
typedef struct position_link {
        TRANSFORM_PTR transform;/* the related transform */
        struct position_link *next;
        struct position_link *inverse;
} POSITION_LINK, *POSITION_LINK_PTR;

typedef struct position_header {
        CHAR *name;                          /* user given name */
        POSITION_LINK_PTR Tnlink;
        POSITION_LINK_PTR Tool_link;
        struct device_info *which_device;    /* related device */
} POSITION, *POSITION_PTR;
```

The purpose of a position equation is to specify in Cartesian space the position and orientation to which the the last link of the robotic device under control should move. To accomplish this task, we solve for the device transform, or actually the T_n transform for that device. For the above position equation, the solution process would entail four homogeneous transform multiplies.

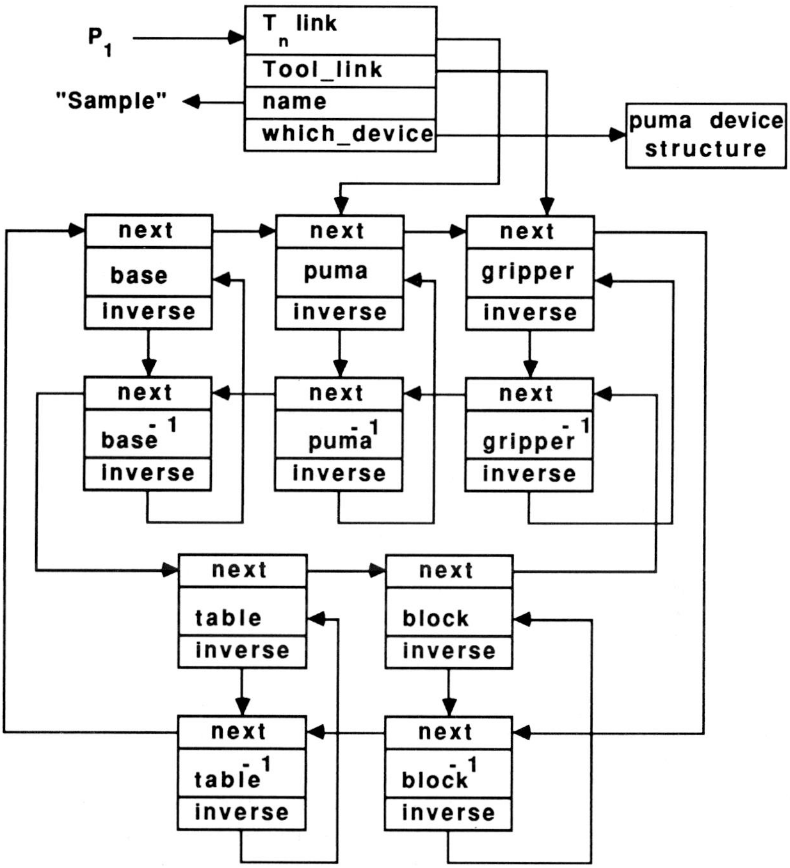

Figure 3. The internal representation of a position equation.

$$[puma] = [base]^{-1}\ [table]\ [block]\ [gripper]^{-1} \tag{8}$$

In our notation, the above procedure for solving for the T_n transform of a device is the $\Gamma()$ function. Its input is a pointer to the desired position equation. This equation is easily solved by using our internal position equation representation. To solve for a particular transform, find the inverse of the transform in the internal representation and follow the next pointers around the loop multiplying the transforms as they are encountered until arriving back at the starting place. For example in Figure 3, start at the $puma^{-1}$ transform and loop through the next fields to arrive at the result in equation 8.

$$T_{P_1} = \Gamma\,(P_1) \tag{9}$$

Here, we ill introduce the shorthand motion, T_{P_1}, meaning the T_n transform for position equation P_1.

Tasks involving coordinated movement between two or more robotic devices can be programmed to have one device in the role of leader or master. This device requires a position specification which does not depend on another device. The follower or slave device uses the T_n transform of the leader to solve for its position and orientation every sample period. Therefore, we make the restriction that only one device may appear in a single position equation without loosing any flexibility of task specification.

To move the desired device to a position we have specified, we pass the position pointer to the move command.

$$move(P_1) \tag{10}$$

The move command puts the position equation onto the *motion queue* of the device listed in the position equation. The motion queue is a first in first out queue which keeps the order of positions for each device to move to.

To specify how the motion is to take place, a motion structure is created to store information pertaining to the move such as velocity, Cartesian motion or joint motion (the specified time to be taken for the movement), segment time, and acceleration time. Motion structures must contains motion information as well as a pointer to the position to which the motion refers.

```
typedef struct motion_info {
        REAL t_seg;                          /* Segment time */
        REAL t_acc;                          /* Acceleration time */
        REAL translational_velocity;
        REAL rotational_velocity;
        INT  Cartesian_motion;               /* TRUE if Cartesian motion */
        INT  segment_time_set;               /* TRUE if user set t_seg */
        POSITION_PTR position_ptr;           /* position equation */
        struct motion_info *next;            /* next motion */
```

```
        CHAR *configuration;                /* use in K^-1 */
        int compute_forward_kinematics;
    } MOTION, *MOTION_PTR;
```

The motion queue is made up of a queue root which contains the head and tail pointers.

```
    typedef struct motion_root_info {
        MOTION_PTR head;
        MOTION_PTR tail;
    } MOTION_ROOT, *MOTION_ROOT_PTR;
```

At this point we are ready to describe the device structure, which must contain variables to keep track of where it is in the task as well as where it is in each position equation execution. Every sample time, the device task list is parsed and in turn each device pointer is passed as an argument to the trajectory generator which generates the next set point for the input device.

```
    typedef struct device_information {
        CHAR *name;                          /* name of device */
        REAL t;                              /* time */
        REAL h;                              /* 0 → 1 between each position */
        INT number_of_axes;
        TRANSFORM_PTR T_n;                   /* the device position */
        TRANSFORM_PTR T_A(-t_acc-τ);         /* T_n at -t_acc-τ */
        TRANSFORM_PTR T_A(-t_acc);           /* T_n at -t_acc */
        TRANSFORM_PTR T_B(-t_acc);           /* T_n of P_B at -t_acc */
        TRANSFORM_PTR T_C(-t_acc-τ);         /* T_n of P_C at -t_acc-τ */
        TRANSFORM_PTR T_C(-t_acc);           /* T_n of P_C at -t_acc */
        TRANSFORM_PTR T_full;                /* for devices with < 6 axes */
        TRANSFORM_PTR D(r);                  /* primary drive transform */
        TRANSFORM_PTR D_s(p);                /* secondary drive transform */
        TRANSFORM_PTR L_T;                   /* for transition equation */
        TRANSFORM_PTR R_T;                   /* for transition equation */
        MOTION_PTR M_B;                      /* motion of current position */
        MOTION_PTR M_C;                      /* motion of next position */
        JOINT_PTR θ_out;                     /* current set point */
        JOINT_PTR θ_last;                    /* last set point */
        JOINT_PTR θ';                        /* current joint velocities */
        JOINT_PTR θ_A(-t_acc);               /* θ at -t_acc */
        JOINT_PTR θ'_A(-t_acc);              /* θ' at -t_acc */
        JOINT_PTR θ_B(-t_acc);               /* θ of P_B at -t_acc */
        JOINT_PTR θ'_B(-t_acc);              /* θ' of P_B at -t_acc */
        JOINT_PTR θ_C(-t_acc);               /* θ of P_C at -t_acc */
        JOINT_PTR θ'_C(-t_acc);              /* θ' of P_C at -t_acc */
        JOINT_PTR θ_BC;                      /* θ_B - θ_C */
        POSITION_PTR P_B;                    /* Position pointer to B */
        POSITION_PTR P_C;                    /* Position pointer to C */
        POSITION_PTR P_new;                  /* New position read in */
        POSITION_PTR P_T;                    /* Transition position */
```

```
    DRIVE_PTR χ_T;                  /* for Cartesian transition */
    DRIVE_PTR χ_A;                  /* drive variables at -t_acc */
    DRIVE_PTR χ'_B(-t_acc);         /* drive velocities of P_B */
    DRIVE_PTR χ'_C(-t_acc);         /* drive velocities of P_C */
    DRIVE_PTR χ_temp;               /* temporary drive variables */
    DRIVE_PTR χ_P;                  /* primary drive variables */
    FUNCTION_PTR K;                 /* forward kinematics */
    FUNCTION_PTR K^-1;              /* inverse kinematics */
    FUNCTION_PTR joint_to_encoder;
    FUNCTION_PTR encoder_to_joint;
    FUNCTION_PTR check_joint_limits;
    FUNCTION_PTR write_servo;
    FUNCTION_PTR read_servo;
    MOTION_ROOT_PTR m_root;         /* root for motion queue */
} DEVICE, *DEVICE_PTR;
```

To allocate space for a device structure we can use a call similar to the one for creating a transform structure. When the motion system is initialized all devices available for use are put on the *system device list*. Before a new device can be added to the system device list, the parameters of the robot must be known, and the robotic device defining functions such as the forward and inverse kinematic functions must be written. To use a device in a given task, it is necessary to transfer it to the *task device list* with the following function call, which returns a pointer to the named device structure. In this case the robotic device is the PUMA manipulator. If more than one PUMA robot were to be controlled, they would require different names to differentiate them on the system device list.

puma = place_on_task_device_list(''PUMA''); (11)

3.0 Trajectory Generation

Trajectory generation for multiple devices essentially consists of trajectory generation for each device separately, treating the other devices as constant transforms each sample time. It is necessary, therefore, to first solve the device which does not depend on the position of the others. We can use the same trajectory generator for every device we wish to coordinate. This trajectory generator needs to be reentrant and to be able to control devices with one to six axis.

We will select an existing trajectory generator and then modify it for multi-device use. There are many trajectory generators published, but if we restrict our search to those which are transform based, generate straight line motion in Cartesian space, and can transition to a moving coordinate frame, the list narrows (Paul 81a) (Taylor 79) (Paul & Zhang 85). We chose (Paul & Zhang 85) for the basis of our trajectory generator because of its versatility, its ease of understanding, and its compact notation. One of its shortcomings, for our purposes, is that they chose to perform all transitions in joint space. We require Cartesian transitions and derive the equations for this. We then augment the trajectory generator for multiple robot control.

3.1 Types of Motions

There are two types of motions which are commonly employed in robotic controllers, joint interpolated motion and straight line or Cartesian motion. Joint interpolated motion moves the controlled device from an initial position to a final position such that all joints start and stop together and the movement in the middle of travel is a constant velocity in each joint space. Joint motion is used to move the device rapidly and to change configurations, but the path of the end effector is unpredictable.

Cartesian motion generates a straight line path in the Cartesian coordinates of the end effector (tool) of the controlled device. With this method, the path of the end effector is precisely specified, but the exact path of each joint is not known.

The motion of a controlled device does not have to be straight line or joint interpolated. For example, a sensor can be used to generate the motion or modify in real time motion generated by the trajectory generator. Or, the device's motion can be specified by a mathematical function, or read from a table of predefined values.

Regardless of the type of motion, the input information is in the form of a homogeneous transform. Position equations made up of transforms are used in the trajectory generator subroutine to generate a path to those positions in the form of a set of joint angles every sample time.

3.2 A Motion Segment

We will start by deriving the equations of motion for a point in α-space which we will call α_{out}. For the moment, the definition of α-space is not important. Later we will show it can be a point in joint space or Cartesian space. We will let a *motion segment* be defined as the straight line from point α_B to point α_C. To move our point along this motion segment, we can use the equation

$$\alpha_{out} = \alpha_B(1-h) + \alpha_C(h) \tag{12}$$

and we will let the parameter h vary from 0 to 1. Or we can write

$$\alpha_{out} = r(\alpha_B - \alpha_C) + \alpha_C \tag{13}$$

and let r = 1 - h as h varies from 0 to 1.

If we let h vary linearly, α_{out} will move from α_B to α_C with a constant velocity. Assuming α_{out} is at rest at α_B, we will define a transition region in which α_{out} starts at rest and accelerates to a constant velocity in time t_{acc}. We will also assume we want the point to start to decelerate time at t_{acc} before arrival at α_C and come to rest at α_C. We will define t_{acc} as the minimum time needed for the robot to accelerate to maximum velocity from rest (see Figure 4).

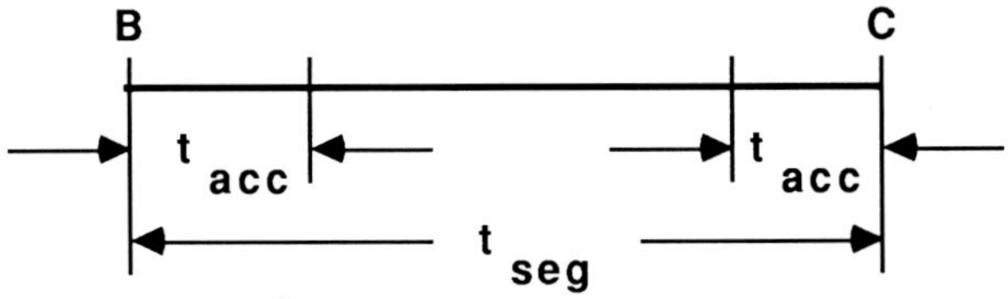

Figure 4. A Motion Segment.

3.3 Combining Multiple Motion Segments

Using our definition of straight motion segments, we can make a path through space by placing them end to end. If we placed two motion segments end to end and labeled the starting point A, the intermediate point B, and the stopping point C, α_{out} would move from rest at point α_A to its eventual goal α_C, stopping completely at point α_B.

3.4 Transitioning Between Two Motion Segments

Although this method of motion specification does allow us to specify a motion task, it does not allow us the flexibility to transition past points without stopping. To eliminate stopping between motion segments, we will start the transition to the next motion segment at the same instant α_{out} is slowing down to stop at the current connecting point. This transition region starts at time t_{acc} before arriving the connecting point and the transition lasts until time t_{acc} past the connecting point. In this way, α_{out} travels with a constant velocity during the middle portions of the line segments and smoothly transitions from motion segment to motion segment in a total time of $2t_{acc}$. We will define the total time for a specified motion segment to be t_{seg}. See Figure 5.

We can view each motion segment as the combination of three types of motions:

(a) Constant velocity motion for t in $[t_{acc}, t_{seg} - t_{acc}]$ (14)

(b) Acceleration from rest for t in $[-t_{acc}, t_{acc}]$ (15)

(c) Deceleration to rest for t in $[t_{seg} - t_{acc}, t_{seg} + t_{acc}]$ (16)

Notice that for adjacent motion segments the deceleration of the previous segment occurs in the same time period as the acceleration of the next segment. Thus, the time intervals

$$[\ t_{seg_{AB}} - t_{acc_{AB}},\ t_{seg_{AB}} + t_{acc_{AB}}\ ] = [\ -t_{acc_{BC}},\ t_{acc_{BC}}\ ] \tag{17}$$

are equal but are measured with respect to two different time axes. The start time of a given motion segment is always zero in the time frame of that segment. For example, for the motion segment from B to C, the zero time point is when time equals $t_{seg_{AB}}$, or

$$t_{0_{BC}} = 0 \ \Big|_{t = t_{seg_{AB}}} \tag{18}$$

We will define a primary motion as the addition of the constant velocity motion (type a) and the acceleration motion (type b). The secondary motion will be defined as the deceleration (type c).

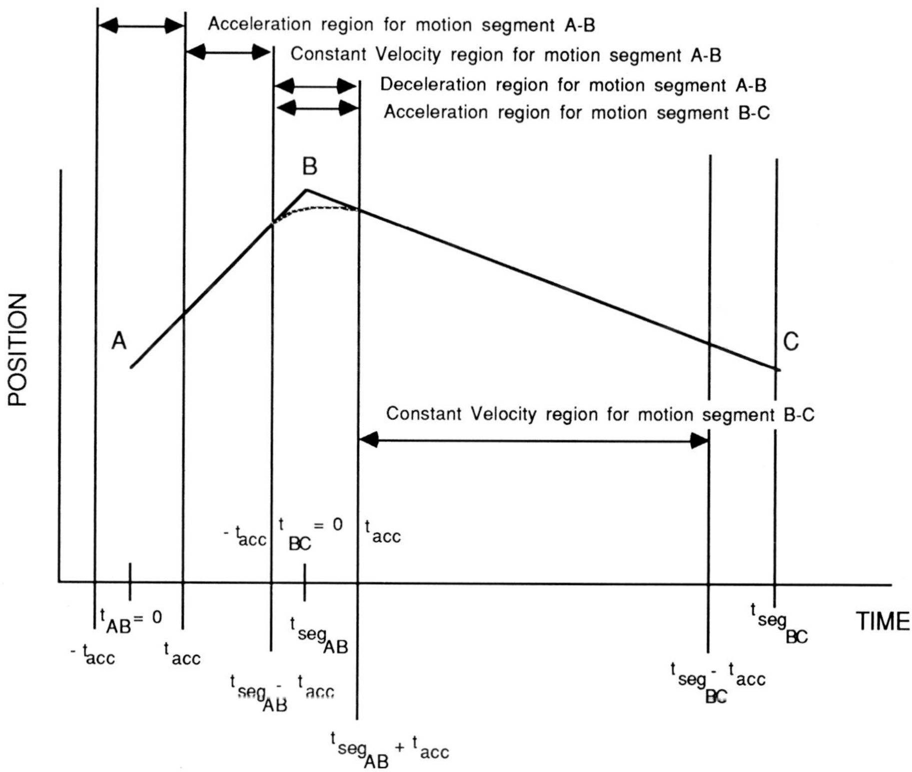

Figure 5. Space trajectory made up of two adjacent motion segments.

3.5 Equations of Motion For The Primary Motion

Looking again at equation (12), we can now derive the the equations for the primary motion.

$$\alpha_{out} = \alpha_B(1 - h) + \alpha_C(h) \tag{12}$$

As h varies from 0 to 1, the motion of α_{out} will move from α_B to α_C. The primary motion, however, is divided into two regions of motion. Thus, we need to determine h(t) as a function of time such that the motion is accomplished both for the acceleration (transition) region and the constant velocity region. We need to match position, velocity, acceleration, and jerk before and after the transition region. Using a seventh order polynomial for h(t) allows us to perform this matching on the transition boundaries. The eight boundary conditions on h(t) are shown in Table 1.

Table 1. Boundary Conditions on h.

time	$-t_{acc}$	t_{acc}	t_{seg}
h	0	t_{acc}/t_{seg}	1
h'	0	$1/t_{seg}$	$1/t_{seg}$
h''	0	0	0
h'''	0	0	0

As we have defined the primary motion as that starting from rest, the boundary conditions at time $t = -t_{acc}$ are all zero. In the region t in $[\, t_{acc},\ t_{seg} - t_{acc}\,]$ we want constant velocity and thus we will use $h(t) = t/t_{seg}$. For this period the time derivative of h is $h'(t) = 1/t_{seg}$.

The parameter h can be expressed as polynomial function of time i.e., h = f (t), where

$$f(t) = a_7t^7 + a_6t^6 + a_5t^5 + a_4t^4 + a_3t^3 + a_2t^2 + a_1t + a_0 \tag{19}$$

in the time interval $-t_{acc} \leq t \leq t_{acc}$.

During the acceleration interval, we let

$$p = \frac{t + t_{acc}}{2\,t_{acc}} \tag{20}$$

then as t varies from $-t_{acc}$ to t_{acc}, p varies from 0 to 1.

Thus the parameter h as a function of p is h = f (p) where,

$$f(p) = a_7p^7 + a_6p^6 + a_5p^5 + a_4p^4 + a_3p^3 + a_2p^2 + a_1p + a_0 \tag{21}$$

with

$$\frac{dp}{dt} = \frac{1}{2\, t_{acc}} \tag{22}$$

Using the first three time derivatives of f(p) and f(p) itself yields four equations. Letting p=0 and 1 at the two boundaries gives eight equations with eight unknowns. Solving for the coefficients yields h = f(p) for the transition region and we get,

for the acceleration region,
$-t_{acc} \le t \le t_{acc}$

$$p = \frac{t + t_{acc}}{2\, t_{acc}}$$

$$h = ((2p-6)p + 5)p^4 \;\; t_{acc}/t_{seg}$$

and, $r = 1 - h.$

In the linear region we have, $t_{seg} - t_{acc} > t > t_{acc}$

$$h = t/t_{seg}$$

and, $r = 1 - h.$

3.6 Equations For The Secondary Motion

As mentioned earlier, the secondary motion represents the deceleration of the previous primary motion segment. Notice that the previous primary motion segment terminates at the time t = $t_{seg} - t_{acc}$. The secondary motion terminates at the time t = t_{acc} (in the time frame of the next segment).

For the second motion during the transition, the motion from $\alpha_{out}(-t_{acc})$ to $\alpha_B(t_{acc})$, we will again use a seventh order polynomial and derive the coefficients using the boundary conditions. Here write $\alpha_B(t_{acc})$ because α_b might be moving. $\alpha_B(t_{acc})$ means where the point α_B is at time t_{acc}. The boundary conditions are listed in Table 2.

Table 2. Boundary Conditions for matching transition.

α	$-t_{acc}$	t_{acc}
α_{out}	$\alpha_{out}-\alpha_B$	0
α'_{out}	$\alpha'_{out}-\alpha'_C$	0
α''_{out}	0	0
α'''_{out}	0	0

Because the secondary motion is added to the primary transition motion described above, we will derive this second transition such that it is the difference between the present position and the next position, with this difference going to zero at t_{acc}. Hence the position boundary condition is $\alpha_{out}-\alpha_B$ at time $t = -t_{acc}$. The velocity boundary condition is required because point C, the target point of the motion segment we are transitioning to, may not be fixed. We can transition to a moving coordinate frame by compensating for the velocity during the secondary motion. Here we could also compensate for moving to an accelerating coordinate frame if it is required. In that case the acceleration boundary condition would need to be $-\alpha''_C$. We will assume we will be transitioning to a constant velocity coordinate frame and thus the acceleration boundary condition is zero.

Taking advantage of the previously defined p variable which changes from 0 to 1 during the transition, we define the polynomial as

$$\alpha = a_7 p^7 + a_6 p^6 + a_5 p^5 + a_4 p^4 + a_3 p^3 + a_2 p^2 + a_1 p + a_0 \tag{23}$$

with

$$p = \frac{t + t_{acc}}{2t_{acc}} \tag{24}$$

Deriving the coefficients using Cramer's rule, we obtain,

$$a_0 = \alpha_{out}(-t_{acc}) - \alpha_B(-t_{acc}) \tag{25}$$

$$a_1 = 2\, t_{acc}\, (\alpha'_{out}(-t_{acc}) - \alpha'_C(-t_{acc})) \tag{26}$$

$$a_4 = -20 a_1 - 35 a_0 \tag{27}$$

$$a_5 = 45 a_1 + 84 a_0 \tag{28}$$

$$a_6 = -36a_1 - 70a_0 \tag{29}$$

$$a_7 = 10a_1 + 20a_0 \tag{30}$$

The subroutine which computes the coefficients of the matching polynominal is called Λ. The time to compute the Λ function is given in Table 3.

Table 3. Trajectory Generator Subroutine Timing.

	multiplies	adds	sine/cosine	atan2	sqrt	time (ms)
Γ(5)	165	120	0	0	0	(2.5)
Δ	119	88	2	3	2	(1.8)
Δ_s	104	88	0	3	2	(1.5)
K^{-1} (puma)	22	14	2	5	1	(2.0)
K (puma)	24	13	4	0	0	(1.2)
D(r)	19	9	1	0	0	(0.4)
$D_s(p)$	64	40	3	0	0	(0.8)
Λ	54	36	0	0	0	(0.7)
misc	45	62	0	0	0	(1.0)

3.7 The Total Motion

Now that we have both motions derived, we can add them to obtain the resultant motion in α-space from $\alpha_{out}(-t_{acc})$ transitioning past α_B and moving toward α_C. The transition past point α_C occurs if another position is specified, and stops at point α_C if no further points are specified. To come to a complete stop using the system just derived, we move to the last position twice.

Our new set point subroutine becomes,

$$
\begin{aligned}
&(-t_{acc} \le t \le t_{acc})\\
&\qquad p = (t + t_{acc})/(2t_{acc})\\
&\qquad h = ((2p - 6)\,p + 5)\,p^4 \;\; t_{acc}/t_{seg}\\
&\qquad r = 1 - h\\
&\qquad \alpha_{out} = [r\,(\alpha_B(-t_{acc}) - \alpha_C(-t_{acc})) + \alpha_C(t)] +\\
&\qquad\qquad ((((a_7 p + a_6)\,p + a_5)\,p + a_4)\,p^3 + a_1)\,p + a_0\\
&(t > t_{acc})\\
&\qquad h = t/t_{seg}\\
&\qquad r = 1 - h\\
&\qquad \alpha_{out} = r(\alpha_B(-t_{acc}) - \alpha_C(-t_{acc})) + \alpha_C(t)
\end{aligned}
\tag{30A}
$$

We can now move a point along a specified path in α-space by moving in a constant velocity in between points and transitioning smoothly from line segment to line segment.

3.8 Various Types of Motion

It is possible to execute motion segments in cartesian or joint space. It is also possible to execute one motion segment in Cartesian space and the next in joint space, or the first in joint and then in Cartesian. We now apply our derivations of the primary and secondary motions along motion segments to joint and Cartesian motion.

3.8.1 Joint Motion

Instead of moving through a series of points in α-space, let us move through a series of points in joint space. The concept is the same, but joint variables are used instead of α variables. A point in joint space is a vector of dimension equal to the degrees of freedom of the device we are trying to control. If the points of the path are given in the form of position equations, which is most convenient for the user, we can convert them into the joint-space of the particular device by using the $\Gamma()$ function (equation 9) and then the inverse kinematic function of the device. We start by listing two position equations which represent positions P_B and P_C in Cartesian space. The actual positions depend on the task and the place in time during the task. For example we might have the following equations.

P_B : [base_of_device][device][gripper] = [reference][pointB] (31)

TOOL: [gripper]

P_C : [base_of_device][device][gripper] = [reference][pointC] (32)

TOOL: [gripper]

To obtain our boundary conditions in joint space, we evaluate the position equations at $-t_{acc}$.

$$T_B(-t_{acc}) = \Gamma(P_B) \tag{33}$$

$$T_C(-t_{acc}) = \Gamma(P_C) \tag{34}$$

Using the inverse kinematic function of the device under control to solve for the joint vector, we get

$$\vec{\theta}_B(-t_{acc}) = \text{device}\rightarrow K^{-1}(T_B(-t_{acc}), \text{configuration}) \tag{35}$$

$$\vec{\theta}_C(-t_{acc}) = \text{device}\rightarrow K^{-1}(T_C(-t_{acc}), \text{configuration}) \tag{36}$$

To obtain the velocity boundary conditions at $-t_{acc}$ listed in Table 2, we evaluate $\vec{\theta}_C$ over two sample periods and take the numerical difference, (the velocity can be also calculated from the Jacobian and the cartesian velocity, if they are known). The current joint velocity is found by subtracting the last two values of $\vec{\theta}_{out}$ and dividing by the trajectory sample time τ. Thus,

$$\vec{\theta}'_C = \left[\vec{\theta}_C(-t_{acc}) - \vec{\theta}_C(-t_{acc}-\tau)\right]/\tau \tag{37}$$

$$\vec{\theta}'_{out} = \left[\vec{\theta}_{out}(-t_{acc}) - \vec{\theta}_{out}(-t_{acc}-\tau)\right]/\tau \tag{38}$$

The equations of motion for a joint move follow that of an α move. The B to C movement is characterized by the equation

$$\vec{\theta}_{out} = r\,(\vec{\theta}_{BC}) + \text{device}\rightarrow K^{-1}(\Gamma(P_C(t)), \text{configuration}) \tag{39}$$

where $\vec{\theta}_{BC} = \vec{\theta}_B(-t_{acc}) - \vec{\theta}_C(-t_{acc})$. The transition equation is the same except joint variables are used instead of α variables. Note that we continuously solve for the P_C position as the transforms defining it might be functionally defined, and this might be moving.

3.8.2 Cartesian Motion

If straight line motion in Cartesian space is desired, we simply use the positions P_B and P_C in Cartesian space. We need to be sure, however, to force the movement to occur in the tool frame of the device.

To move from P_B to P_C in Cartesian space, we will introduce a drive transform (Paul 81a) into the position equation. The drive transform represents a translational and an orientation change from P_B to P_C, both proportional to r.

Any orientation change between two positions can be composed of a rotation ψ around a unit vector, $\hat{u}$. If we equate the rotation around a unit vector with the three Euler rotations,

$$\text{Rot}(\hat{u}, \psi) = \text{Euler}(\phi, \theta, \psi) = \text{Rot}(z, \phi)\,\text{Rot}(y, \theta)\,\text{Rot}(z, \psi) \tag{40}$$

we see that the first two Euler angles, ϕ and θ rotated around z and y axis, can be used to line up the unit vector in such a manner so that ψ yields the equivalent rotation. Multiplying the first two Euler angle gives,

$$\text{Rot}(z,\ \phi)\ \text{Rot}(y,\ \theta) = \begin{bmatrix} \cos\phi\cos\theta & -\sin\phi & \cos\phi\sin\theta & 0 \\ \sin\phi\cos\theta & \cos\phi & \sin\phi\sin\theta & 0 \\ -\sin\theta & 0 & \cos\theta & 0 \\ 0 & 0 & 0 & 1 \end{bmatrix} \tag{41}$$

The third Euler angle, ψ, rotates around the z axis, and therefore we want the unit vector to be the new z axis defined by the first two Euler angles. Using the third column of the above matrix product, we have:

$$\hat{u} = \cos(\phi)\sin(\theta)\hat{i} + \sin(\phi)\sin(\theta)\hat{j} + \cos(\theta)\hat{k} \tag{42}$$

where $0 \le \theta < \pi$, and $\hat{i}$, $\hat{j}$, and $\hat{k}$ are unit vectors along the x, y, and z axes of the frame with respect to which the rotation is performed. The translation of x,y,z is simply the linear difference between the two reference frames, and can be written as:

$$\text{translation} = \text{rx}\ \hat{i} + \text{ry}\ \hat{j} + \text{rz}\ \hat{k}\ , \tag{43}$$

Here r is a changing parameter for $1 \rightarrow 0$.

If we want to find the drive transform required to move from position P_B to position P_C, we use the elements of the homogeneous transform which describe P_B with respect to P_C. Using the following relationships we can derive values which functionally define the drive transform.

To find the Euler angles, we use the formulas found in (Paul & Zhang 85):

$$\psi = \tan^{-1} + \frac{\sqrt{(n_y - o_x)^2 + (a_x - n_z)^2 + (o_z - a_y)^2}}{(n_x + o_y + a_z - 1)} \tag{44}$$

$$\theta = \tan^{-1} + \frac{\sqrt{(a_x - n_z)^2 + (o_z - a_y)^2}}{(n_y - o_x)} \tag{45}$$

and as $0 \le \theta < \pi$ and $0 \le \psi < \pi$ we may obtain ϕ uniquely as

$$\phi = \tan^{-1} \frac{(a_x - n_z)}{(o_z - a_y)} \tag{46}$$

The set of variables x, y, z, ϕ, θ, ψ will be referred to as the drive variables given by χ, and the corresponding data structure contains these REAL variables.

```
typedef struct drive_variables {
        REAL  x, y, z, φ, θ, ψ;
} DRIVE, *DRIVE_PTR;
```

The general form of the drive transform is:

$$D(r) = \text{Trans}(rx,\ ry,\ rz)\ \text{Rot}(\hat{u},\ r\psi) \tag{47}$$

To use the drive transform to move in Cartesian space, we need to modify equation P_C to include the drive transform. For equation (32), for example, the internal position equation would be changed to include the drive transform, D(r).

P_C : [base_of_device][device][gripper] = (48)

[reference][pointC][D(r)]

TOOL: [gripper]

To derive the symbolic form of the drive transform, we will use the procedure described in (Paul & Zhang 85). To solve for the drive transform, we rewrite the position equations as the concatenation of three transforms, L, D(r), and R. Transform L is the concatenation of those transforms to the left of the drive transform, transform R is the concatenation of those transforms to the right of the drive transform, and D(r) represents the drive transform. It is important to place the drive transform in the position equation so that the Cartesian movement occurs in the tool frame of the device.

Rewriting P_C to include the drive transform results in

$$P_C : T_C = L_C\ D(0)\ R_C \tag{49}$$

where D(0) represents the identity transform.

As r varies from 1 to 0 during the move, we want the Cartesian point to move from P_B to P_C. Therefore, when r = 1, (at time $-t_{acc}$), we are at the start of the motion and the device position and orientation at $T_B(-t_{acc})$ is related to P_C by the following equation.

$$T_B(-t_{acc}) = L_C\,(-t_{acc})\,D(1)\,R_C(-t_{acc}) \tag{50}$$

which can be solved for D(1), as

$$D(1) = L_C^{-1}\,(-t_{acc})\,T_B(-t_{acc})\,R_C^{-1}\,(-t_{acc}) \tag{51}$$

To solve for the drive parameters x,y,z,ϕ, θ, and ψ, we set

$$L_C^{-1}\,(-t_{acc})\,T_B(-t_{acc})\,R_C^{-1}\,(-t_{acc}) = \begin{bmatrix} n_x & o_x & a_x & p_x \\ n_y & o_y & a_y & p_y \\ n_z & o_z & a_z & p_z \\ 0 & 0 & 0 & 1 \end{bmatrix} \tag{52}$$

and use equations (43) through (46). The function which solves for the drive parameters is the Δ function. Thus,

$$\chi_P = \Delta(T_B(-t_{acc}),\, L_C,\, R_C)\,. \tag{53}$$

To move in Cartesian motion we use the equation

$$T_n = \Gamma\,(L_C(t)\,D(r)\,R_C(t)) \tag{54}$$

and let r vary from 1 to 0. To generate the position vector in joint space, we call the inverse kinematics for the device.

$$\vec{\theta}_{out} = \text{device}{\rightarrow}K^{-1}\,(\Gamma(L_C(t)\,D(r)\,R_C(t)),\ \text{configuration}) \tag{55}$$

Cartesian transitions were not derived in (Paul & Zhang 85). Because we feel it is a necessity, we will derive a method for Cartesian transitions.

When we transition from one Cartesian motion path segment to another, it is necessary that the resulting movement is predictable in Cartesian space. If we were to calculate the transition in joint space, unpredictable tool motion would result for the brief period of the transition, which for many application purposes is unacceptable. The transitions from Cartesian motion and joint motion into joint motion can occur in joint space however, because the motion of the tool is not specified in joint motion unpredictable tool motions results.

To obtain the position boundary conditions listed in Table 2 in Cartesian space, we will calculate the drive parameters and find the difference between two successive drive parameters to obtain the velocities. To compute the secondary drive parameters we use the subroutine Δ_s which computes the Euler angles given in T_n, and the concatenated transforms to the left and right of the drive transform. Thus to compute the drive parameters for the transition region, we call:

$$\chi_T = \Delta_s(T_n(-t_{acc}), L_T, R_T)\,. \tag{56}$$

where,

$$L_T = L_B\ (-t_{acc})\,, \tag{57}$$

$$R_T = R_B\ (-t_{acc})\,. \tag{58}$$

To find the derivatives of the drive parameters we first calculate

$$\chi_{temp} = \Delta_s(T_n(-t_{acc}-\tau), L_T, R_T) \tag{59}$$

and then take the difference and divide by τ.

$$\chi'_T = \left[\chi_T - \chi_{temp}\right]/\tau \tag{60}$$

To calculate the derivatives of the drive parameter for P_C we use,

$$\chi'_C = \left[\Delta_s(T_C(-t_{acc}-\tau), L_C(-t_{acc}), R_C(-t_{acc}))\right]/\tau\,. \tag{61}$$

Because P_B might involve a functionally defined transform in its definition, we form a new position equation, P_T, used only for Cartesian transitions.

$$P_T : T_n = L_T\, D_s(p)\, R_T \tag{62}$$

where,

$$L_T = L_B\ (-t_{acc})\,, \tag{63}$$

and, $R_T = R_B\ (-t_{acc})\,,$ (64)

and $D_s(p)$ is the secondary drive transform. To use this drive transform requires that all drive variables are in the same form. For this reason we will not use the rotation around a vector, but will instead use the drive variables as they are computed, in the form of Euler angles. The form of the secondary drive parameter calculation subroutine has less computations required than the primary drive parameter subroutine described above. To compute the secondary drive transform we use the p variable during the transition region, where

$$p = \frac{(t + t_{acc})}{2\, t_{acc}} \tag{65}$$

During the transition, the drive parameters are computed using the seventh order polynomial whose coefficients were found using the drive variable boundary conditions. The form of the a coefficients are shown in equations 25-30.

$$x_p = ((((a_{x_7} p + a_{x_6})\, p + a_{x_5})\, p + a_{x_4})\, p^3 + a_{x_1})\, p + a_{x_0} \tag{66}$$

$$y_p = ((((a_{y_7} p + a_{y_6})\, p + a_{y_5})\, p + a_{y_4})\, p^3 + a_{y_1})\, p + a_{y_0} \tag{67}$$

$$z_p = ((((a_{z_7} p + a_{z_6})\, p + a_{z_5})\, p + a_{z_4})\, p^3 + a_{z_1})\, p + a_{z_0} \tag{68}$$

$$\phi_p = ((((a_{\phi_7} p + a_{\phi_6})\, p + a_{\phi_5})\, p + a_{\phi_4})\, p^3 + a_{\phi_1})\, p + a_{\phi_0} \tag{69}$$

$$\theta_p = ((((a_{\theta_7} p + a_{\theta_6})\, p + a_{\theta_5})\, p + a_{\theta_4})\, p^3 + a_{\theta_1})\, p + a_{\theta_0} \tag{70}$$

$$\psi_p = ((((a_{\psi_7} p + a_{\psi_6})\, p + a_{\psi_5})\, p + a_{\psi_4})\, p^3 + a_{\psi_1})\, p + a_{\psi_0} \tag{71}$$

The drive transform $D_S(p)$ has the general form of a Euler angle frame description,

$$D_s(p) = \begin{bmatrix} ? & -C(\phi_p)C(\theta_p)S(\psi_p) - S(\phi_p)C(\psi_p) & C(\phi_p)S(\theta_p) & x_p \\ ? & -S(\phi_p)C(\theta_p)S(\psi_p) + C(\phi_p)C(\psi_p) & S(\phi_p)S(\theta_p) & y_p \\ ? & S(\theta_p)S(\psi_p) & C(\theta_p) & z_p \\ 0 & 0 & 0 & 1 \end{bmatrix}, \tag{72}$$

where the **n** vector (represented as "?" in equation (72)) can be determined by $\mathbf{n} = \mathbf{o} \times \mathbf{a}$.

If we evaluate the transition position equation we obtain the straight line path which drives the robotic device from point $P_B(-t_{acc})$ to $P_B(t_{acc})$. But what we actually need is the difference in Cartesian space which will be added to the first equation of motion. This amount which matches the difference in Cartesian space is given by the relationship:

$$[T_{matching}]^{-1} [T_B(-t_{acc})] \tag{73}$$

where,

$$T_{matching} = \Gamma(L_T \, D_s(p) \, R_T) \tag{74}$$

3.9 Summary of Motion Equations

By deriving the equations of motion as two motions added together, we obtain a very simple method to smoothly transition between functionally defined positions. The two types of motions, joint and Cartesian, result in four types of transition possibilities: joint motion to joint motion, joint motion to Cartesian motion, Cartesian motion to joint motion, and Cartesian motion to Cartesian motion. However, because we chose to transition in joint space when moving to a joint space move, we have three unique equations. The four transition cases are summarized below.

Joint motion to joint motion

$$\vec{\theta}_{out} = r(\vec{\theta}_{BC}) + \tag{75}$$

$$\text{device}\rightarrow K^{-1}(\Gamma(P_C(t)), \text{configuration})$$

$$+ ((((a_7 p + a_6)\, p + a_5)\, p + a_4)\, p^3 + a_1)\, p + a_0$$

where the coefficients of the polynominal are determined in joint space.

Joint motion to Cartesian motion

$$\vec{\theta}_{out} = \text{device}\rightarrow K^{-1} \; (\Gamma(L_C(t)\, D(r)\, R_C(t)), \text{configuration}) \tag{76}$$

$$+ ((((a_7\, p + a_6)\, p + a_5)\, p + a_4)\, p^3 + a_1)\, p + a_0$$

Cartesian motion to Cartesian motion

$$T_n - \Gamma(L_C(t)\, D(r)\, R_C(t))) \tag{77}$$

$$T_{matching} = \Gamma(L_T \, D_s(p) \, R_T)) \tag{78}$$

$$T_n = T_n \; [T_{matching}]^{-1} [T_B(-t_{acc})] \tag{79}$$

$$\vec{\theta}_{out} = \text{device}\rightarrow K^{-1}\ (T_n,\ \text{configuration}) \tag{80}$$

Cartesian motion to joint motion

$$\vec{\theta}_{out} = r(\vec{\theta}_{BC}) + \text{device}\rightarrow K^{-1}\ (\Gamma(P_C(t)),\ \text{configuration})$$

$$+ ((((a_7\ p + a_6)\ p + a_5)\ p + a_4)\ p^3 + a_1)\ p + a_0 \tag{81}$$

3.10 Integration of the Trajectory Generator

The motion control system is divided into two types of functions, those related to the trajectory generator, and those called by the user, in a *user_task* (), to specify the robot task. The number of *user_task* ()'s written correspond to how the user wants to break up the communication coordination of the task. Global variables in the user file allows semaphores to be set up for intertask communication.

The trajectory generator is run through entirely each sample time for each device on the device task list. The *user_task* ()'s are time sliced and run for a brief period each sample time to give the illusion of parallel execution on a serial machine. Figure 6 illustrates the tasks which require execution each sample time for a master / slave robot task with two *user_task* ()'s. Equations (82) through (84) describe a possible master /slave coordination specification.

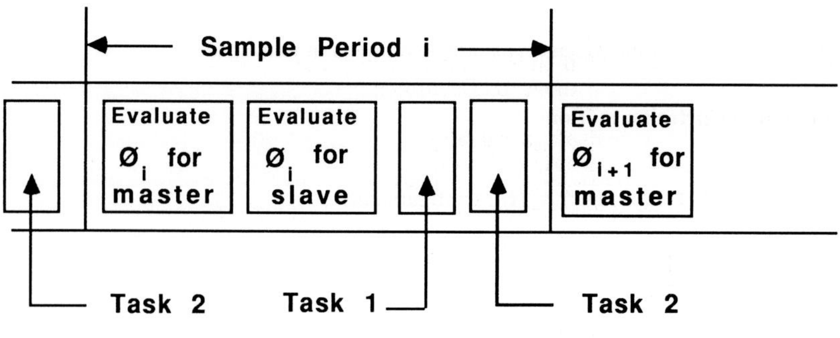

Figure 6. Task execution during the *i*th sample period.

3.10.1 Device Independent Trajectory Generator

The most common mode of operation of the trajectory generator is to call compute_current_set_point (), which executes the equations of motion derived earlier. A less common, although more computational expensive, mode occurs during a transition from one position to the next. At the start of a transition the trajectory generator function computes the coefficients of motion to be used in compute_next_set_point (). This computation is spread over three (or multiple) sample times to lower the computation demand which occurs at $(t_{seg}-t_{acc})$. The new position P_{new} is read in at $(t_{seg}-t_{acc}-\tau)$, motion information gathered at $(t_{seg}-t_{acc})$ and the transition polynominal coefficients are computed at $(t_{seg}-t_{acc}+\tau)$. The following function executes a device independent trajectory generator.

```
trajectory_generator (d)
DEVICE_PTR d;
{
        /*
         * check to see if this sample period is
         * one before the start of the next transition
         */
        if (d→t = d→M_C→t_seg − d→M_C→t_acc − τ)
        {
                compute_current_set_point (d)
                /* get the next position to move to */
                d→M_B = d→M_C
                d→M_C = dequeue_motion (d)
                if (d→M_C ≡ NULL)
                {
                        /*
                         * if no more motions on queue
                         * move to last position.
                         */
                        d→P_new = d→P_C
                        d→M_C = d→M_B
                        d→M_C→t_acc = 3.0*τ
                        /* max latency to start new position is 6.0*τ */
                        d→M_C→t_seg = 6.0*τ
                }
                else
                {
                        d→P_new = d→M_C→position_ptr
                }
                h_temp = d→h
                d→h = 0.0
                /* execute the first part of equation 61 */
                d→T_C(−t_acc − τ) = Γ(d→P_new)
                d→h = h_temp
                if ((d→M_B→Cartesian_motion ≡ TRUE) &&
                        (d→M_C→Cartesian_motion ≡ TRUE))
                {
                        /* transition in Cartesian space */
```

```
                d→T_A(−t_acc − τ) = d→T_full
                d→Cartesian_transition = TRUE
        }
        else
        {
                /* transition in joint space */
                d→θ⃗_C(−t_acc − τ) =
                        d→K^-1(d→T_C(−t_acc − τ),
                                d→M_C→configuration)
                d→Cartesian_transition = FALSE
        }
}
/*
 * check to see if we are at the start of the transition
 */
if (d→t = d→M_B→t_seg − d→M_B→t_acc)
{
        compute_current_set_point (d)
        h_temp = d→h
        d→h = 1.0
        d→T_B(−t_acc) = Γ(d→P_C)
        d→L_new(−t_acc) = L_new
        d→R_new(−t_acc) = R_new
        if (d→Cartesian_transition ≡ TRUE)
        {
                /*
                 * Save parameters for transitioning in
                 * Cartesian space
                 */
                d→L_C(−t_acc) = L_C
                d→R_C(−t_acc) = R_C
                d→T_A(−t_acc) = d→T_full
                d→χ_T = Δ_s(d→T_A(−t_acc), d→L_C(−t_acc), d→R_C(−t_acc))
                d→χ_A = Δ_s(d→T_A(−t_acc−τ), d→L_C, d→R_C(−t_acc))
                d→χ'_A(−t_acc) = (d→χ_T − d→χ_A) / τ
                d→χ'_C(−t_acc) = Δ_s(d→T_C(−t_acc − τ),
                        d→L_new(−t_acc), R_new(−t_acc))
                d→χ'_C(−t_acc) /= τ
                /* fill in transition equation */
                d→L_T = L_B(−t_acc)
                d→R_T = R_B(−t_acc)
        }
        else
        {
                /*
                 * Save parameters to transition in
                 * joint space
                 */
                d→θ⃗_B(−t_acc) = d→K^-1(d→T_B(−t_acc),
                        d→M_B→configuration)
                d→θ⃗_A(−t_acc) = d→θ⃗_out
                d→θ⃗'_A(−t_acc) = d→θ⃗'_out
                d→h = 0.0
```

```
            d→T_C(−t_acc) = Γ(d→P_new)
            d→θ⃗_C(−t_acc) = d→K^−1(d→T_C(−t_acc),
                  d→M_C→configuration)
            d→θ⃗′_C(−t_acc) = (d→θ⃗_C(−t_acc) − d→θ⃗_C(−t_acc − τ)) / τ
        }
}
/*
 * check to see if we are at the sample period after
 * the start of the transition
 */
if (d→t = d→M_B→t_seg − d→M_B→t_acc + τ)
{
        d→P_B = d→P_C
        d→P_C = d→P_new
        if (d→M_B→Cartesian_motion ≡ TRUE)
        {
                remove_drive_transform (d→P_B)
        }
        if (d→Cartesian_transition ≡ TRUE)
        {
                /*
                 * calculate drive transition parameters
                 * equations 66 - 71
                 */
                a_0 = d→χ_T
                a_1 = (d→χ′_A(−t_acc) − d→χ′_C(−t_acc)) *
                        2(d→M_C→t_acc)
                a_4 = −20a_1 − 35a_0
                a_5 = 45a_1 + 84a_0
                a_6 = −36a_1 − 70a_0
                a_7 = 10a_1 + 20a_0
        }
        else
        {
                /*
                 * calculate joint transition variables
                 * equations 25 - 30 in joint space
                 */
                a_0 = d→θ⃗_A(−t_acc) − d→θ⃗_B(−t_acc)
                a_1 = (d→θ⃗′_A(−t_acc) − d→θ⃗′_C(−t_acc)) *
                        2(d→M_C→t_acc)
                a_4 = −20a_1 − 35a_0
                a_5 = 45a_1 + 84a_0
                a_6 = −36a_1 − 70a_0
                a_7 = 10a_1 + 20a_0
        }
        /* calculate the primary drive parameters */
        d→χ_P = Δ(d→T_B(−t_acc), d→L_new(−t_acc), d→R_new(−t_acc))
        if (d→M_C→Cartesian_motion ≡ TRUE)
        {
                install_drive_transform (d→P_C)
        }
        else
```

```
            {
                    d→θ⃗_BC = d→θ⃗_B(−t_acc) − d→θ⃗_C(−t_acc)
            }
            if ( d→M_C→segment_time_set ≡ FALSE)
            {
                    /* compute t_seg from velocities */
                    d→M_C→t_seg = MAX (
                            d→χ_P→ψ / d→M_C→rotational_velocity,
                            ((d→χ_P→x)^2 + (d→χ_P→y)^2 + (d→χ_P→z)^2)^½
                            ------------------------------------------- )
                                  d→M_C→translational_velocity
            }
            /* reset the time counter */
            d→t = −d→M_C→t_acc + τ
            compute_next_set_point (d)
        }
        /* increment the time counter by the sample period */
        d→t += τ
}
```

The subroutine compute_next_set_point () executes the algorithm derived in the section 3.7. It first checks to see if the input device is in a transition region. If it is, p and h are computed and the during_transition flag is set. The appropriate motion equation is selected, either joint, equation (39), or Cartesian, equation (55), and $\vec{\theta}_{out}$ is computed. If the device is in a transition region, the appropriate matching equation of motion is selected depending on the mode of the movement.

Notice that because we control devices with less than six degrees of freedom, we save the result of the Γ() function in T_{full}. This is used in the trajectory generator to allow path planning to occur in a higher dimension than the degrees of freedom of the device we are controlling. At the end of the function, if the user wishes to take take advantage of this option, the forward kinematic function is called to ensure the Cartesian position and orientation of a device with less that six axes represents the true device position. The compute_forward_kinematics flag is contained in the MOTION structure. The forward kinematics function is always called if the move is in joint motion to allow tracking of all robotic devices under control.

```
compute_current_set_point (d)
DEVICE_PTR d;
{
        d→θ⃗_last = d→θ⃗_out
        /*
         * check to see if we are at the start of
         * the transition period
         */
        if (d→t = d→M_B→t_seg − d→M_B→t_acc)
        {
                if (d→Cartesian_transition ≡ FALSE)
                {
```

```
                /*
                 * Approximate joint vector to reduce computation
                 * spike. Don't approximate Cartesian to
                 * Cartesian transition.
                 */
                d→θ⃗out += d→θ⃗′ * τ
                d→Tn = d→K(d→θ⃗out)
                d→Tfull = d→Tn
                d→Tn^-1 = [d→Tn]^-1
                d→θ⃗′ = (d→θ⃗out − d→θ⃗last) / τ
                d→check_joint_limits(θout)
                d→joint_to_encoders(d)
                return
        }
}
/*
 * Execute the primary motion.
 */
if (d→t ≤ d→MC→tacc)
{
        p = (d→t + d→MC→tacc) / 2(d→MC→tacc)
        d→h = ((2p − 6)p + 5)p^4 d→MC→tacc / d→MC→tseg
        during_transition = TRUE
}
else
{
        d→h = d→t / d→MC→tseg
        during_transition = FALSE
}
r = 1. - d→h
if (d→MC→Cartesian_motion ≡ TRUE)
{
        /* equation 77 */
        d→Tn = Γ(d→PC(with D(r)))
        if (during_transition && d→Cartesian_transition ≡ TRUE)
        {
                /* equation 78 and 79 */
                d→Tmatching = Γ(d→PT(with Ds(p)))
                d→Tn *= [d→Tmatching]^-1 d→TB(−tacc)
        }
        d→Tfull = d→Tn
        d→θ⃗out = d→K^-1(d→Tn,d→MC→configuration)
        if (d→MC→compute_forward_kinematics ≡ TRUE)
        {
                d→Tn = d→K(d→θ⃗out)
        }
        if (during_transition && d→Cartesian_transition ≡ FALSE)
        {
                /* equation 76 */
                d→θ⃗out +=
                        ((((a7p + a6)p + a5)p + a4)p^3 + a1)p + a0
                d→Tn = d→K(d→θ⃗out)
        }
```

```
        }
        else
        {
                /* equation 75 */
                d→θ⃗out = r(d→θBC) +
                        d→K^-1(Γ(d→PC(without D(r))),
                                d→MC→configuration)
                if (during_transition ≡ TRUE)
                {
                        d→θ⃗out += ((((a7p + a6)p + a5)p + a4)p^3 + a1)p + a0
                }
                d→Tn = d→K(d→θ⃗out)
        }
        d→Tn^-1 = [d→Tn]^-1
        d→θ⃗' = (d→θ⃗out − d→θ⃗last) / τ
        d→check_joint_limits(θout)
        d→joint_to_encoders(d)
}
```

3.10.2 Overall System Design

Given the trajectory_generator () and compute_current_set_point () functions, we can write the main () function for the trajectory generator system with the help of some real time operating system calls. Although an interrupt structure could be used to schedule the real time tasks in the system, a real time operating system works nicely and keeps the code relatively portable given the real time executive exists for the target processor. The function of the real-time executive is explained later in the chapter.

```
main ()
{
        while (1)
        {
                wait_for_new_sample_time ();
                current_device = device_task_list→head;
                while (current_device != NULL)
                {
                        current_device→read_servo ();
                        current_device→write_servo ();
                        current_device = current_device→next;
                }
                current_device = device_task_list→head;
                while (current_device != NULL)
                {
                        trajectory_generator (current_device);
                        current_device = current_device→next;
                }
                time_slice (user_task1, user_task2, ...);
        }
}
```

The read_servo () and write_servo () functions "read from" and "write to" servo processors which perform the appropriate control of the corresponding robotic device joints.

4.0 Application Examples

The specification of a robot task involving more than one robotic device is now easy to accomplish given the above method of position equation specification. To illustrate motion coordination, we will give two examples of a master/slave arrangement. First we describe a robotic task which performs non-destructive testing of composite materials using two five axes motion devices with a common axis and a one axis table. The master five axes device holds an ultrasonic transmitter and the slave device holds the receiver. As a second example, we will describe the position equations needed to perform a down-hand weld using a two axes positioning table and a six axes robot. Lastly we will augment the scanning task to show how communication coordination between multiple devices can be implemented.

4.1 Example One: Non-Destructive Testing

Non-destructive testing of composite materials involves scanning the surface of the composite material with an ultrasonic transmitter while tracking the scanning motion on the opposite side with an ultrasonic receiver, see Figure 2. Variations in the ultrasonic signal produce a internal map of the part including any defects as the part is scanned.

We will assume we are scanning boat hulls to locate any defects in the hull material. Two five axes robots sharing a common axis and a one axis table will be used to produce the scanning motion for the sensors. The robotic device master_device will hold the transmitter and the device slave_device will hold the receiver. The boat hull will be mounted on the table. The [surface] transform will be functionally defined and will return the next scanning surface point each sample time. The position equation which specifies table motion is given by:

$$P_{table} : [table] = [table\_orientor] \tag{82}$$

TOOL:[table]

where the [table_orientor] transform evaluates the [surface] transform and extracts the required table information. The position equation to control the master_device is

$$P_{transmit} : [base\_of\_master][master\_device][transmitter] = \tag{83}$$

$$[table \rightarrow T_n][boat\_hull][surface]$$

TOOL : [transmitter]

And the equation for the slave_device is

$$P_{receive} : \text{[base\_of\_slave][slave\_device][receiver]} = \text{[base\_of\_master][master\_device}\rightarrow T_n\text{]} \text{[transmitter][thickness\_of\_boat\_hull]} \quad \text{TOOL : [receiver]} \tag{84}$$

By first moving the table, and master and slave manipulators into their starting locations:

$$P_{table\_start} : \text{[table]} = \text{[table\_start]} \quad \text{TOOL:[table]} \tag{85}$$

$$P_{master\_start} : \text{[base\_of\_master][master\_device][transmitter]} = \text{[table}\rightarrow T_n\text{][boat\_hull][scan\_start\_point]} \quad \text{TOOL : [transmitter]} \tag{86}$$

$$P_{slave\_start} : \text{[base\_of\_slave][slave\_device][receiver]} = \text{[bench][boat\_hull][scan\_start\_point][thickness\_of\_boat\_hull]} \quad \text{TOOL : [receiver]} \tag{87}$$

$$\text{move}(P_{table\_start}); \tag{88}$$

$$\text{move}(P_{master\_start}); \tag{89}$$

$$\text{move}(P_{slave\_start}); \tag{90}$$

we can accomplish the test with three calls to move:

$$\text{move}(P_{table}); \tag{91}$$

$$\text{move}(P_{transmit}); \tag{92}$$

$$\text{move}(P_{receive}); \tag{93}$$

4.2 Example Two: Down-hand Welding

In a down hand welding system, the weld must occur on the part so that the weld puddle does not roll off the part. This can be accomplished by forcing the z axis of the weld path coordinate frame to be aligned with the gravity vector at the current weld point. Often, a two axis table, upon which the part to be welded is mounted, and a six axis robot, carrying the weld torch, are used to perform the weld (similar to Figure 1). Although this system is redundant, as only five degrees of freedom are needed to perform the weld or six if rotational control around the torch is required, it is often used for convenience.

To control a system of this type require two basic position equations. The first position equation aligns the z coordinate axis of the weld frame with the gravity vector, and the second position equation locates the weld torch at the welding point. Thus the position equation to control the positioning table to align the weld vector down hand might be

$$P_{dh} : [\text{table\_base}][\text{py\_table}][\text{part}][\text{weld\_path}(h)] = [\text{down\_hand}] \qquad (94)$$

TOOL: [weld_path(h)]

where py_table is the device pointer to a two axis table, [table_base] is the transform describing the location of the py_table in the workspace, [weld_path(h)] is a parametric definition of the weld path ($0 \le h \le 1$), and [down_hand] is defined as

$$[\text{down\_hand}] = \begin{bmatrix} C_\delta & -S_\delta & 0 & 0 \\ S_\delta & C_\delta & 0 & 0 \\ 0 & 0 & 1 & 0 \\ 0 & 0 & 0 & 1 \end{bmatrix} \qquad (95)$$

where the function associated with the [down_hand] transform solves for the table orientation angle δ every sample time. Notice that because positional information is not used in the py_table K^{-1} () function, we don't need to include it in [down_hand].

The second position equation specifies the robot to move to the weld point, which depends on the position and orientation of the table, thus

$$P_{weld} : [\text{robot\_base}][\text{welding\_robot}][\text{weld\_torch}] = \qquad (96)$$

$$[\text{table\_base}][\text{py\_table}\rightarrow T_n][\text{weld\_path}(h)]$$

TOOL: [weld_torch]

where [py_table$\rightarrow T_n$] is treated as a constant transform each sample time.

After moving to the starting point of the weld

$$P_{table\_start} : [table\_base][py\_table][part][weld\_path(0)] = [down\_hand(0)] \quad (97)$$

$$\text{TOOL: } [weld\_path(0)]$$

$$P_{start\_weld} : [robot\_base][welding\_robot][weld\_torch] = \quad (98)$$

$$[table\_base][py\_table \rightarrow T_n][weld\_path(h)]$$

$$\text{TOOL: } [weld\_path(0)]$$

For the above example to work correctly, each sample time we need to first evaluate the py_table to determine the next down hand position and then evaluate the puma device to correctly place the weld torch. The evaluation order is defined by the order the user puts the devices on the device task list.

As a final example, let us expand on the previous non-destructive testing example. We will add a gantry robot to supply the scanning robots with new boat hulls. Using two user_task() subroutines, we can easily separate the functionality of the robotic tasks while providing communication coordination. The pseudo code below outlines the two concurrent tasks.

```
/*
 * An example of two user-written tasks which provide
 * communication coordination between three robotic
 * devices while dividing the functionality of
 * the task for ease of programming.
 */
static int in_position = FALSE;
static int testing_done = FALSE;
static int all_done = FALSE;

user_task1 ()
{
        DEVICE_PTR gantry;
        int i;

        gantry = place_on_task_device_list ("GANTRY");
        /* have the gantry supply boat hulls */
        for (i = 0; i < number_of_boat_hulls; i++)
        {
                pick_up_new_boat_hull (gantry);
                place_on_table (gantry);
                in_position = TRUE;
                if (i == number_of_boat_hulls -1)
                        all_done = TRUE;
                while (testing_done == FALSE)
                        /* wait */ ;
```

```
                remove_boat_hull (gantry);
        }
        all_done = TRUE;
}

user_task2 ()
{
        DEVICE_PTR table, master_device, slave_device;
        int i;

        table = place_on_task_device_list ("TABLE");
        master_device = place_on_task_device_list ("MASTER");
        slave_device = place_on_task_device_list ("SLAVE");
        while (all_done == FALSE)
        {
                /* scan the boat hull */
                while (in_position == FALSE)
                        /* wait */ ;
                .
                .
                .
                move (P_table_start);
                move (P_master_start);
                move (P_slave_start);
                .
                move (P_table);
                move (P_transmit);
                move (P_receive);
                .
                .
                .
                testing_done = TRUE;
                in_position = FALSE;
        }
}
```

5.0 Trajectory Computational Issues

The main functions which consume the majority of the time in terms of arithmetic operations during trajectory calculations are shown in Table 3. Using these numbers as a guideline for estimating computation times on other processors, we performed bench mark tests of these functions on a 25MHz Motorola 68020 processor with cache enabled and with a 20MHz Motorola 68881 co-processor. These times are shown in the last column. The $\Gamma(5)$ function computations assume five transform multiplies. The misc entry combines joints to encoders conversion, joint limit checks, and calculations involving the **r** parameter.

Tables 4 through 7 show the time requirements for each transition combination of trajectory generation. The number in each category is the number of times the function is called followed by the time required in milliseconds to execute the function on the above processor.

Given a uniform sample time the point at $t_{seg} - t_{acc}$ during a Cartesian to Cartesian transition requires the most computation and defines the sample time. During this sample period, two Γ () functions are required, one K^{-1} (), one primary drive transform and one secondary drive transform. This computation is similar to Paul's (Paul 81a) in which the worst case sample period is the transition from joint motion to Cartesian motion. During this period, the current motion must be converted to Cartesian space, requiring first a Γ () function then a K () solution. Following this two drive transforms, coefficients of the matching polynomial, and an K^{-1} () solution are required. The worst case occurs during a transition to Cartesian motion (Paul & Zhang 85). This corresponds to a Γ () function call, two K^{-1} () solutions, and a drive transform calculation.

By implementing the trajectory generator to transition over three sample periods, the computation spike at $t_{seg} - t_{acc}$ can be lowered. In our trajectory generator we allow tracking in joint motion which Paul (Paul 81a) does not allow, and we transition in Cartesian space which Paul (Paul & Zhang 85) do not allow.

The actual sample time used depends on the task, the degree of complexity of any functionally defined transforms, and the number of robotic devices which are being controlled at one time.

Table 4. Cartesian to Cartesian Transition
Trajectory Generator Set Point Time Requirements.

	linear region	normal transition	$t_{seg} - t_{acc} - \tau$	$t_{seg} - t_{acc}$	$t_{seg} - t_{acc} + \tau$
Γ	1 (2.5)	2 (5.0)	2 (5.0)	2 (5.0)	2 (5.0)
Δ	0	0	0	0	1 (1.8)
Δ_s	0	0	0	3 (4.5)	0
K^{-1}	1 (2.0)	1 (2.0)	1 (2.0)	1 (2.0)	1 (2.0)
K	0	0	0	0	0
D(r)	1 (0.4)	1 (0.4)	1 (0.4)	1 (0.4)	1 (0.4)
D_s(p)	0	1 (1.2)	0	0	1 (1.2)
Λ	0	0	0	0	1 (0.7)
misc	(1.0)	(1.0)	(1.0)	(1.0)	(1.0)
totals	(5.9)	(8.2)	(8.4)	(12.9)	(11.7)

Table 5. Cartesian to Joint Transition
Trajectory Generator Set Point Time Requirements.

	linear region	normal transition	$t_{seg} - t_{acc} - \tau$	$t_{seg} - t_{acc}$	$t_{seg} - t_{acc} + \tau$
Γ	1 (2.5)	1 (2.5)	2 (5.0)	2 (5.0)	1 (2.5)
Δ	0	0	0	0	1 (1.8)
Δ_s	0	0	0	0	0
K^{-1}	1 (2.0)	1 (2.0)	1 (2.0)	2 (4.0)	1 (2.0)
K	1 (1.2)	1 (1.2)	0	1 (1.2)	1 (1.2)
D(r)	0	0	1 (0.4)	0	0
$D_s(p)$	0	0	0	0	0
Λ	0	0	0	0	1 (0.7)
misc	(1.0)	(1.0)	(1.0)	(1.0)	(1.0)
τ	(6.7)	(6.7)	(8.2)	(11.2)	(11.2)

Table 6. Joint to Cartesian Transition
Trajectory Generator Set Point Time Requirements.

	linear region	normal transition	$t_{seg} - t_{acc} - \tau$	$t_{seg} - t_{acc}$	$t_{seg} - t_{acc} + \tau$
Γ	1 (2.5)	1 (2.5)	2 (5.0)	2 (5.0)	1 (2.5)
Δ	0	0	0	0	1 (1.8)
Δ_s	0	0	0	0	0
K^{-1}	1 (2.0)	1 (2.0)	2 (4.0)	2 (4.0)	1 (2.0)
K	0	1 (1.2)	1 (1.2)	1 (1.2)	1 (1.2)
D(r)	1 (0.4)	1 (0.4)	0	0	1 (0.4)
$D_s(p)$	0	0	0	0	0
Λ	0	0	0	0	1 (0.7)
misc	(1.0)	(1.0)	(1.0)	(1.0)	(1.0)
τ	(5.9)	(7.1)	(11.2)	(11.2)	(9.6)

Table 7. Joint to Joint Transition
Trajectory Generator Set Point Time Requirements.

	linear region	normal transition	$t_{seg} - t_{acc} - \tau$	$t_{seg} - t_{acc}$	$t_{seg} - t_{acc} + \tau$
Γ	1 (2.5)	1 (2.5)	2 (5.0)	2 (5.0)	1 (2.5)
Δ	0	0	0	0	1 (1.8)
Δ_s	0	0	0	0	0
K^{-1}	1 (2.0)	1 (2.0)	2 (4.0)	2 (4.0)	1 (2.0)
K	1 (1.2)	1 (1.2)	1 (1.2)	1 (1.2)	1 (1.2)
D(r)	0	0	0	0	0
$D_s(p)$	0	0	0	0	0
Λ	0	0	0	0	1 (0.7)
misc	(1.0)	(1.0)	(1.0)	(1.0)	(1.0)
totals	(6.7)	(6.7)	(11.2)	(11.2)	(11.2)

6.0 Real-Time Trajectory Computation in a Multiprocessor System

In a multiprocessor system three new requirements arise which are not encountered in a uniprocessor design: first, each problem must be partitioned into tasks; second, each task must be scheduled for execution on one or more processors; third, synchronization of control and data flow must be performed during execution (Gajski & Peir 85). Many multiprocessor robot controllers attack the problem of robot control based on the parallel execution of the equations describing the robot dynamics (Luh & Lin 82) (Nigam & Lee 85) (Kasahara & Narita 85) (Liu & Chen 86). None of these controllers, however, discuss positional control and path generation issues. The Cartesian path of the end effector is assumed to be generated elsewhere and the joint variables $\mathbf{q}$, $\dot{\mathbf{q}}$, and $\ddot{\mathbf{q}}$, describing this path are input variables to their controllers. The objective of our multiprocessor system is to efficiently compute these joint variables for multiple robotic devices based on Cartesian relationships in the robot task and pass these variables to servo processors which implement the appropriate dynamic control, see Figure 7.

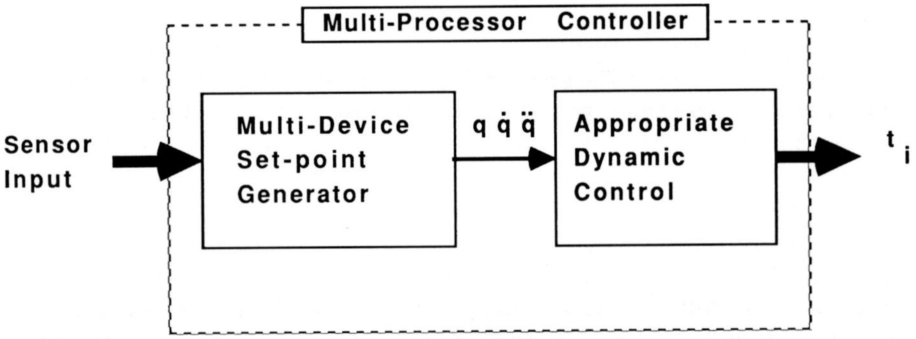

Figure 7. A block view of the multiprocessor controller.

Recently (Zhang 86) designed a multiprocessor robot controller that computes Cartesian position / force motions, as well as control dynamics very efficiently. The controller, however, is designed to control one six axes robot manipulator with a closed form inverse kinematic solution. It is our goal to design a controller capable of generating joint variable positional information for coordinating multiple robotic devices of various degrees of freedom, even if the inverse kinematic functions are iterative in nature. Ahmad and Li (Ahmad & Li 89) also designed a multiprocessor system for calculating the entire control calculation needed for a single robot control. Their design consisted of a single main processor scheduling a number of sub processors or arithmetic processors. Their calculations show approximately six processors are necessary for optimal processor versus speed throughput.

The dynamic equations for robot manipulators are well known (Paul 81) (Bejczy 74) (Luh et.al 80). The equations can be determined from the Lagrange-Euler for formulation or from the Newton-Euler formulation. It has been determined that Newton-Euler formulation lead's to efficient and fast recursive on line dynamic computation, as Newton-Euler schemes lead to O(n) computations where Lagrange-Euler schemes typically lead to $O(n^4)$ computations, here n is the number of joints. Once a computation scheme is selected, an optimizing procedure can be used to find the lowest time computation schedule. Luh and Lin (Luh & Lin 82) use a "variable" branch and bound method, while Nigam and Lee (Nigam and Lee 85) use the recursive nature of the Newton-Euler equations of motion to optimally fit the computations on a pipelined architecture. Kasahara and Narita (Kasahara & Narita 85) developed a general offline approach to multiprocessor scheduling by developing a depth first/implicit heuristic search (DF/IHS) algorithm which can be used to find a near optimal schedule for any series of computational tasks to be run on a multiprocessor system.

Unlike dynamic formulations, however, the computations required by our controller depend on the robot task for which the controller is programmed. In addition, some tasks involving sensor input or function calculation execute in a variable amount of time depending on the state of the environment. Even in a task with little sensor input, a robotic device's position might depend on the solution of another device such as in a master-slave relationship. For these reasons, it is unpractical to preschedule the calculations. We need a technique which will partition the tasks quickly and easily, and an algorithm which will schedule and synchronize the tasks in real-time. We want to accomplish this for a variety of robot tasks given a variable number of processors.

We will give a preference to a design which satisfies our design goals of an efficient programming and development environment, and no special computer architectures. We would like to use off the shelf hardware whenever possible to reduce development time and design costs. We want the final design to be flexible enough to allow the user to add processors as the task demands increase, with very little change to the system. The processor structure should be transparent to the motion control system, as we want an easy upgrade to the next generation of hardware and software.

6.1 Hardware Organization

Before examining the task partioning and scheduling algorithm, we will first outline the multiprocessor target controller. The term multiprocessor is quite broad and can encompass many computer configurations. For our discussion here, we restrict our controller to consist of a host and a tightly-coupled multiprocessor computer, see Figure 8. The host, which can be most any machine the user wishes, will be used to write and compile the robot motion programs. The executable code is down-loaded to the tightly-coupled system for execution.

To make the host more valuable to the entire robot controller, we link the host to the tightly-coupled system with an an Ethernet connection (Intel, DEC, & Xerox 82). This will allow the host to participate in the execution of the motion control system, either as a simulator which receives joints angles and plots them, or as a sensor processor which provides input to modify the robot trajectory in real-time.

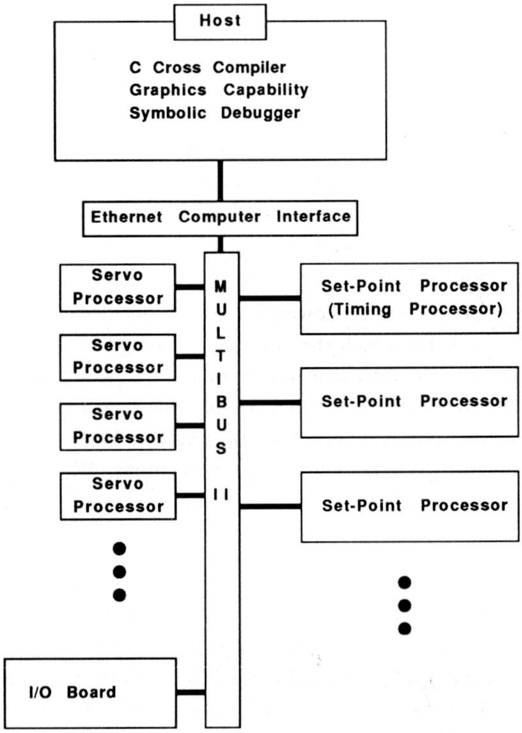

Figure 8. The multiprocessor controller with host.

The tightly-coupled system consists of multiple processors on a common bus with the processors divided into two types dependent on the functionality, set-point and servo processors. Set-point processors will be used to compute the joint angle set points for each motion device. The user can select the number of set-point processors depending on the computational demands of the task. Servo processors will digitally servo the joints to the commanded position as computed by the set-point processors. Servo processors execute an appropriate control algorithm dependent on the dynamic formulation selected. The number of servo controlled joints determines the number of servo processors that are to be utilized

We assume that each set-point processor board has 2 Megabytes of local memory and a real-time operating system on-board. In addition, 64K Bytes of dual-ported memory is available on each board and is used for synchronization and communication among the multiple processes, see Figure 9. Each processor can keep its own status of the system in its shared memory with other processors updating it as they complete various sub tasks. One set-point processor will be the timing processor, and will contain the master clock of the system.

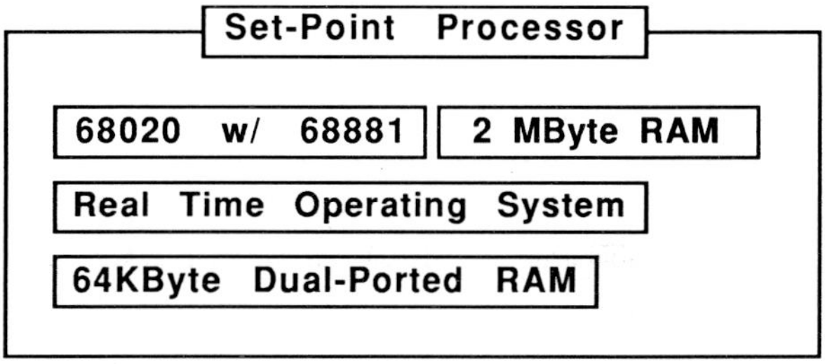

Figure 9. The components of a set-point processor.

6.2 System Organization

The multiprocessor system function is organized into four functional units; the user interface, the motion library, the partitioning and scheduling routines, and the real-time executive. The relationship between these are shown in Figure 10.

The user interface contains user written functions of two types, *user_task* ()'s and homogeneous transform *defining functions*. The *user_task* () functions describe the motion generation and coordination relationships between the various robotic devices for the given task by using homogeneous transform based position equations. As the *user_task* ()'s are executed the robotic motion operations are performed. Homogeneous transforms requiring sensor input or mathematical functions to describe them are input parameters to the *defining function* calls. These function calls can either compute the values directly or access other special sensor processors over the bus. The *user_task* () functions makes calls to the motion library. The library is divided into two types of functions, user-callable, and those involved with trajectory generation. User-callable functions are used to build position equations, construct motion queues, and describe types of motion. Trajectory generation is accomplished by reading the user created motion queues and performing coordinated motion for the related devices.

The partitioning and scheduling functions are divided into three; fill_in_TTL (), update (), and the set-point algorithm. The fill_in_TTL () function is called by the trajectory generator function and is used to partition the tasks each sample time. The update () routine is called by the user written *defining functions* for synchronizing the tasks. The fill_in_TTL () and update () functions will be described later in greater detail. The set-point algorithm schedules and synchronizes in real-time the execution of the trajectory generation process for all robotic devices involved as well as the user interface tasks. The set-point algorithm is developed in further sections and is the main focus of this chapter.

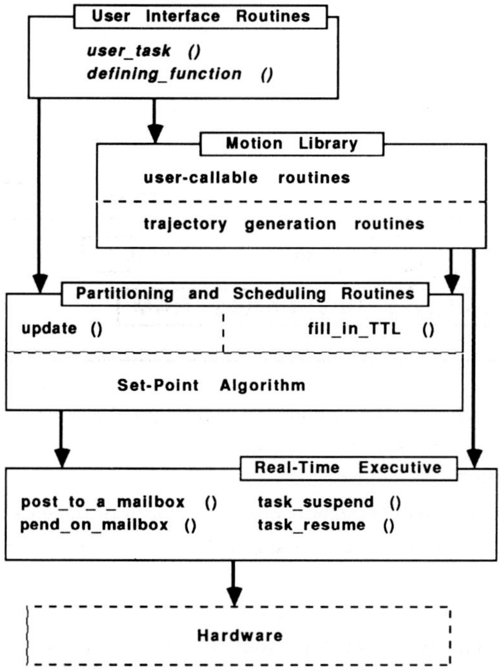

Figure 10. The functional divisions in the multiprocessor controller.

To aid in task scheduling and synchronization, the set-point algorithm makes calls to a real-time executive resident on each processor board. A real-time executive is a hardware dependent program which interfaces the scheduling code to hardware in a transparent way. We will use the real-time executive for three basic purposes.

First, we will take advantage of the intertask communication and synchronization capabilities inherent in the real-time operating system to schedule event driven tasks. An event occurs when a mailbox (memory location) is either read from or written to. A processor pending on a mailbox wakes up when an event writes a 1 to the mailbox. We will assume our real-time executive contains the commands; post_to_a_mailbox (at_this_address), which sets the memory location to a 1, and pend_on_mailbox (at_this_address), which causes the processor to wait for an event to fill the mailbox. By locating the mailbox for each processor in its own shared RAM, these primatives allow intertask communication between the processors with a minimum of bus access. A processor pending on an event does not need to access the bus to check its own mailbox contents.

Second, we want the ability to suspend a task running on one processor, and restart it on another processor. The real-time executive gives us this capability, using the commands task_suspend () and task_resume (). When a task is suspended, the current state of the task is stored in a Task Control Block (TCB). This contains information such as the registers and status bytes of the processor at the time of suspension. Also stored, along with the TCB, are the registers and status bytes of the coprocessor. The local variables of all the subroutines entered up to the point of suspension are stored in a block of memory we will call the function stack-space. The total memory required to store the TCB, the floating point registers, and the function stackspace for one task is 2K bytes. This assumes no large arrays of local variables will be declared. If a large array is needed, it can be declared as a global variable.

Third, when a task is suspended, the 2K bytes of status information is stored in shared RAM on the timing processor board. When a processor other than the timing processor is in a state to resume execution of the task, the 2K bytes of status information must be transferred over the system bus to the shared RAM of the second processor. The real-time executive can then resume this task as if it suspended it. Global variable information must also be transferred.

In a system with shared memory, two or more processes can have access to the same stored variable. For purposes of synchronization and to ensure that one process does not modify a shared data structure while another process is using it, an ability to lock and unlock shared memory must exist. The locking mechanism is simply a memory location in shared RAM associated with the data structure we wish to lock. Locking sets the variable to a one. Unlocking sets the variable to a zero.

The lock command needs to be implemented with a test and set assembly code instruction. In this way, reading a lock returns the state of the lock (0 or 1) and then automatically sets it to a 1, thus locking the lock. This operation is indivisible, or atomic.

The user created software is written on the host processor, compiled, linked to the motion control library and to the fill_in_TTL () and update () routines, and then down-loaded to the multiprocessor system. A copy of the set-point algorithm and the real-time executive exist in ROM on every set-point processor board. When the executable motion control code is down-loaded into the tightly-coupled system, a copy is placed in local memory on each set-point processor board. The shared memory on each board is used for task synchronization once the task is initiated.

7.0 Task Partitioning for Real-Time Execution

Before task partitioning can take place, we must first identify the tasks to be executed in the system. For the system to be useful, this must be done in a general fashion as the goal is to obtain a flexible system which can coordinate multiple robotic devices. The basis for motion specification, independent of robotic application, is the transform based position equation. We will therefore start with a simple position equation and find a general method for listing the tasks we can partition.

For example, let us form a position equation such as the following,

$$(P_{sensor})\text{:[robot\_base][robot][gripper]=[conveyor][block]} \tag{99}$$

TOOL:[gripper]

and instruct the robot to move to that position.

$$\text{move}(P_{sensor}); \tag{100}$$

In so doing, we expect the robot to move to the location defined by the homogeneous transforms. If the transforms change value during the motion, we expect to see this change reflected in the motion of the robot. If a transform value does change, it must have an associated *defining function* which computes its value each set-point. If a transform value does not change during a motion, the *defining function* is a dummy function and is not evaluated. The end result, however, is that each sample period, each transform in the current position equation obtains its final value through the evaluation of its *defining function.* The dummy function is a special case of this more global view.

As an example of this concept, let us look at the above position equation, P_{sensor}. The conveyor position might be calculated mathematically and the block position might depend on a sensor reading. Whatever the form of input, a process is necessary to determine its value each set-point. Similarly, the robot's T_n transform has as its *defining function* the trajectory generator described in section 2.2. Thus, from a conceptual viewpoint, the evaluation of all homogeneous transforms currently being used in the system defines the evaluation of all required tasks every sample period. Trajectory generation is accomplished as a function of the T_n transform it modifies, making the set-point generation for all devices a natural result of updating all changing Cartesian transforms.

From a task point of view, the Cartesian task can be partitioned into subtasks which consist of updating each transform in the current position equation such that the position equality holds. Partitioning the Cartesian task involves first evaluating each transform, as each might be functionally defined, and then multiplying them together.

In our system, we partition the subtasks in the form of a linked list of transforms to be evaluated every sample time called the Transform Task List (TTL). Figure 11 shows the TTL when the P_{sensor} position equation is active.

At the end of every sample time, the trajectory generator of each active robotic device puts the transforms which need to be evaluated for its motion generation next sample time on the TTL. This is accomplished with the function fill_in_TTL (). The order each device adds transforms to the TTL follows the priority order on the *task device list*. The user sets this order in the *user_task* (). A copy of the transforms in the TTL are also put in shared memory on the other processors. These are used to test the completion status of these transforms.

A dummy transform entitled *user* is created for each *user_task* () and is put on the end of the TTL. When a *user* transform is executed from the TTL, the *user_task* () subroutine is run for a selectable short period of time. We will use 2.0 milliseconds in our examples.

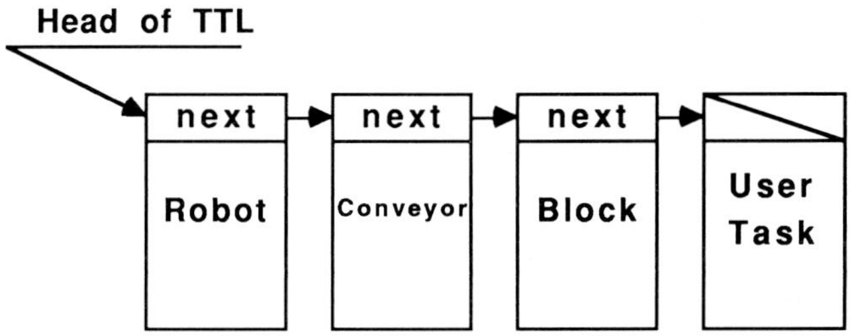

Figure 11. The Transform Task List for the P_{sensor} position equation.

8.0 Task Scheduling and Synchronization (The Set-Point Algorithm)

Given that we have a linked list of transform tasks to be evaluated every sample period, we now develop a simple set-point algorithm to schedule the tasks in real-time. The basic flow of this algorithm is shown in Figure 12. The set-point algorithm consists of four fundamental states; pend, search, fetch, and execute. In the pend state, the processor is waiting for a wake-up call (pending on a mailbox) to start processing. The search state consists of searching the TTL for available transform tasks to run. In the fetch state, the processor transfers the data needed to perform the task, and finally the task is executed in the execute state.

This algorithm runs on every set-point processor available. The timing processor receives an external wake-up call from the master clock to start the sample time. The timing processor then wakes all other available processors.

8.1 Pend State

Every processor contains a mailbox variable which is initialized to zero. When an event occurs, such as a timer or another processor writing a one to that location, the processor wakes from its pending state and is ready to enter the search state. At the start of every sample time, the shared memory on the timing processor board contains the Transform Task List for the current sample period, as put there by the active device's trajectory generator functions. Also contained in shared memory are the global variables needed by the transform tasks. The memory map of the shared memory is shown in Figure 13.

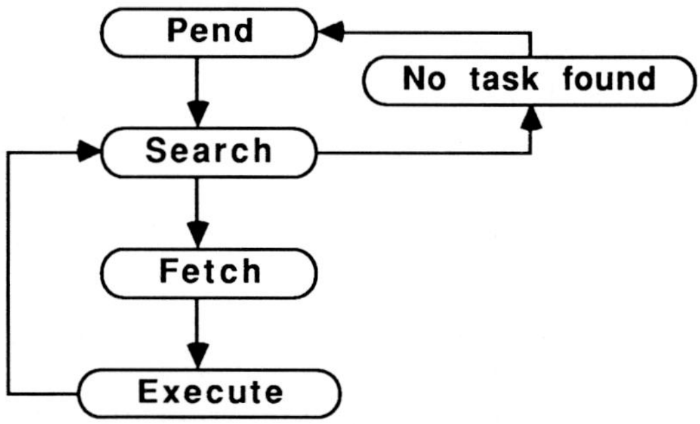

Figure 12. The set-point algorithm overview.

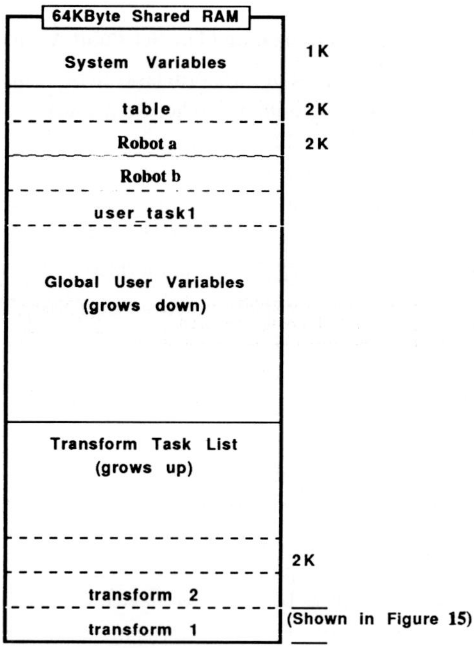

Figure 13. The memory map of shared RAM.

8.2 Search State

In the search state, Figure 14, the processor searches the TTL to find the next available transform task for execution. To do this it must first gain access to the system bus, for the TTL is located in the shared memory on the timing processor.

Each set-point processor board is given a bus access priority. If a processor has control of the bus, it retains control even if another processor wishes to use the bus. If more than one processor tries to access the bus at the same time, the processor with the higher priority gains access.

Even though the TTL is located in the shared RAM on the timing processor, the timing processor must gain access to the bus when searching the TTL in order to lockout any other processor from searching the TTL. Once the bus access has been granted, each transform task is checked on the list. If a transform task is found which is not suspended, and is not currently being executed, the task can be executed and the processor proceeds to the fetch state. If no transform task is found, the system bus is released and the processor goes into the pend state.

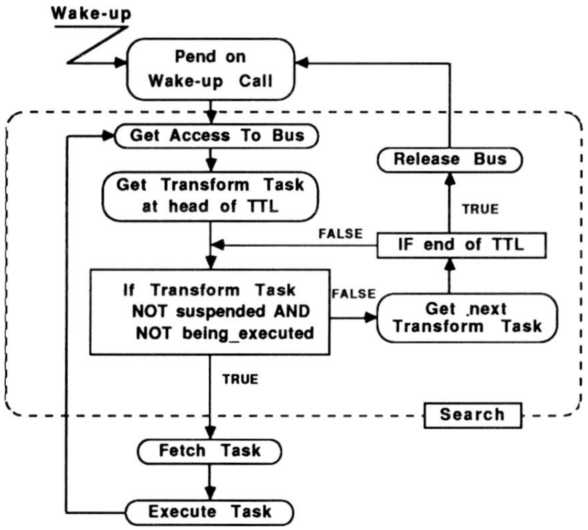

Figure 14. The search state of the set-point algorithm.

8.3 Fetch State

When a task is suspended under the real-time operating system, the information describing the state of the task is stored in a Task Control Block. This contains information such as register values and status bytes which enable the process to correctly resume execution. The floating point registers of the coprocessor are similarly stored and the local variables are saved in a stack. This information, as well as variables used by the set-point algorithm, are stored with the transform on the TTL, see Figure 15. When a processor wishes to run a task, all this information, as well as any global variables, must be first read by the processor.

Upon entering the fetch state, Figure 16, the processor first sets the *being_executed* flag to true, preventing future processors from running the task. If the task had previously been suspended, it is necessary to transfer the TCB, the floating point registers and the function stackspace to the shared RAM of the fetching processor. If the task has not previously been suspended, there is no reason to transfer this information as it does not contain any information. Independent of the suspension status of the task, the global variables must be transferred to shared RAM. An exception to this transfer is when the processor accessing the TTL in the timing processor as this information is already in its shared RAM.

After the pertinent information has been transferred, the processor releases the bus and moves into the execution state.

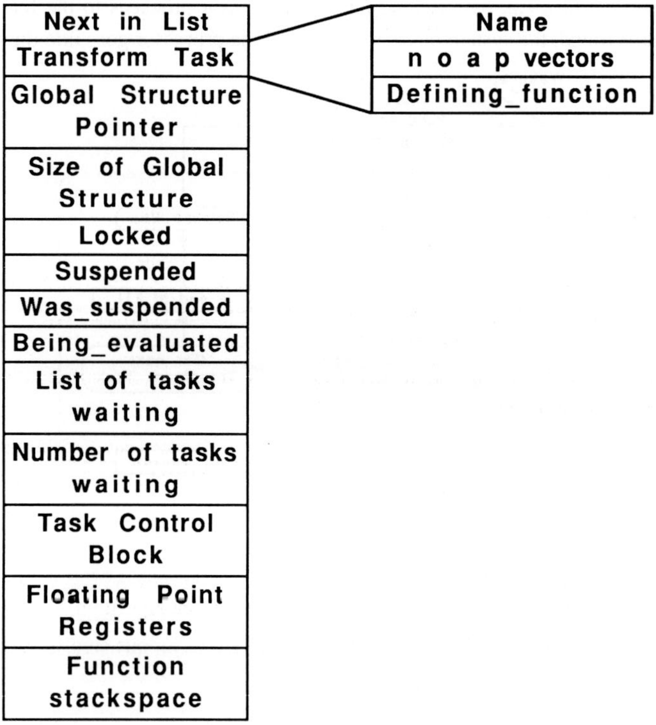

Figure 15. The variables associated with a transform task.

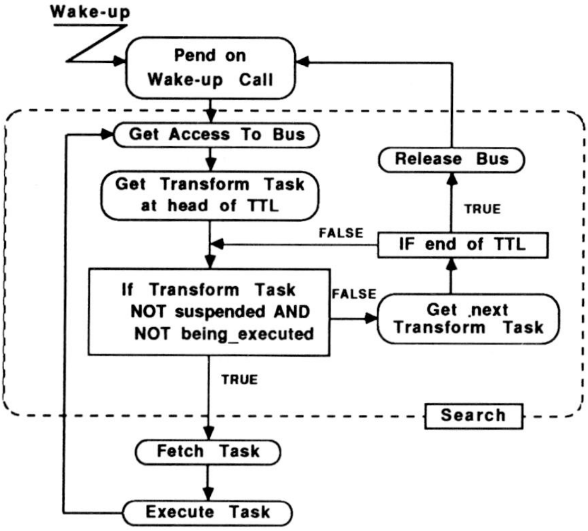

Figure 16. The fetch state of the set-point algorithm.

8.4 Execution State

In the execution state, the *defining_function* of the transform task is executed. Although this function updates the contents of the transform, it does not have to perform all the calculations involved. For example, in a transform task involving vision processing, the function can be written to extract the information from a vision board on the bus or through an Ethernet connection. In this way, the execution time of the *defining_function* can be kept low and fairly constant. The bus traffic can be minimal because only the transform information needs to be extracted from the vision board.

When the *defining_function* is run, three different outcomes are possible; the function can run to completion, the function can be suspended because it depends on a transform task not yet completed, or the function can unlock its parent transform because it has computed its contents for the current sample period.

If the task runs to completion, Figure 17a, access to the system bus is obtained and the updated global variables are sent back to shared RAM on the timing processor. If the task is not a *user_task* (), the transform task is removed from the TTL. The bus control is maintained and the search state is entered.

If the function needs to use a transform which is functionally defined, it checks its completion state by looking in the list of transforms in its shared memory. These transforms are a copy of the ones in the TTL and are unlocked as their *defining_functions* compute their contents. If a transform which is needed for further execution is locked, Figure 17b, the task is suspended, control of the bus is obtained, and a request is put in to unsuspend this task when the dependent transform becomes unlocked. The task status information as well as the global variable information is read back into the shared memory on the timing processor and the processor enters the search state.

When a user writes a *defining_function*() for a transform, it is necessary to call the function *update*() when the transform computation has been completed. *Update* () is a function available to the user as a library call. It is compiled along with motion control functions on the host. When *update*() is executed, it first obtains control of the system bus. It then writes the transform's **n**, **o**, **a**, and **p**, vectors to all copies of that transform in each processor's shared RAM, and correspondingly unlocks each copy of the transform, see Figure 17c. It then searches the TTL and unsuspends any transform tasks waiting on its completion. It then wakes up the same number of processors as number of tasks it just unsuspended or as many processors that are available in a pend state. It then releases control of the bus and continues execution.

Note that a user function will never update the parent transform, as it is a dummy transform.

8.5 The Set-Point Algorithm

The set-point algorithm in its entirety is now listed below.

Step 1: Wait for a wake-up call from another processor. In the case of the timing processor, receive an interrupt from the system timer. When a wake-up call is received, go to step 2.

Step 2: Get access to the system bus. Start at the head of the TTL and get the first transform task from the list. Go to step 3.

Step 3: If the *being* executed flag is true, go to step 4, else go to step 5.

Step 4: Get the next transform task off the TTL and go to step 3. If no more tasks exist go to step 1.

Step 5: Set the *being_executed* flag to TRUE. If the processor accessing the bus is the timing processor, release the bus and go to step 6, as the timing processor will run with the task related variables in its shared memory. If the processor is not the timing processor and the task has previously been suspended, load the Task Control Block, the floating point registers, and the function stack space into its own shared memory. Regardless of the state of the task, load the global variables into shared memory. Release the bus and go to step 6.

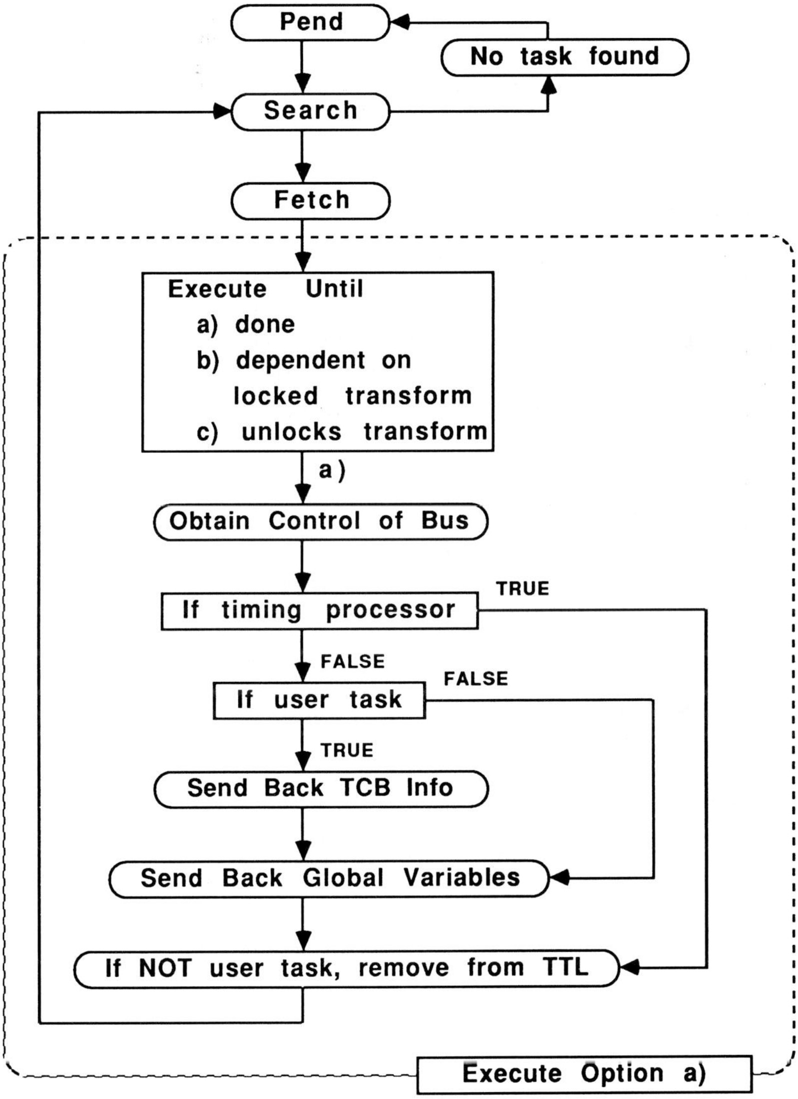

Figure 17a. The a) option of the execute state of the set-point algorithm.

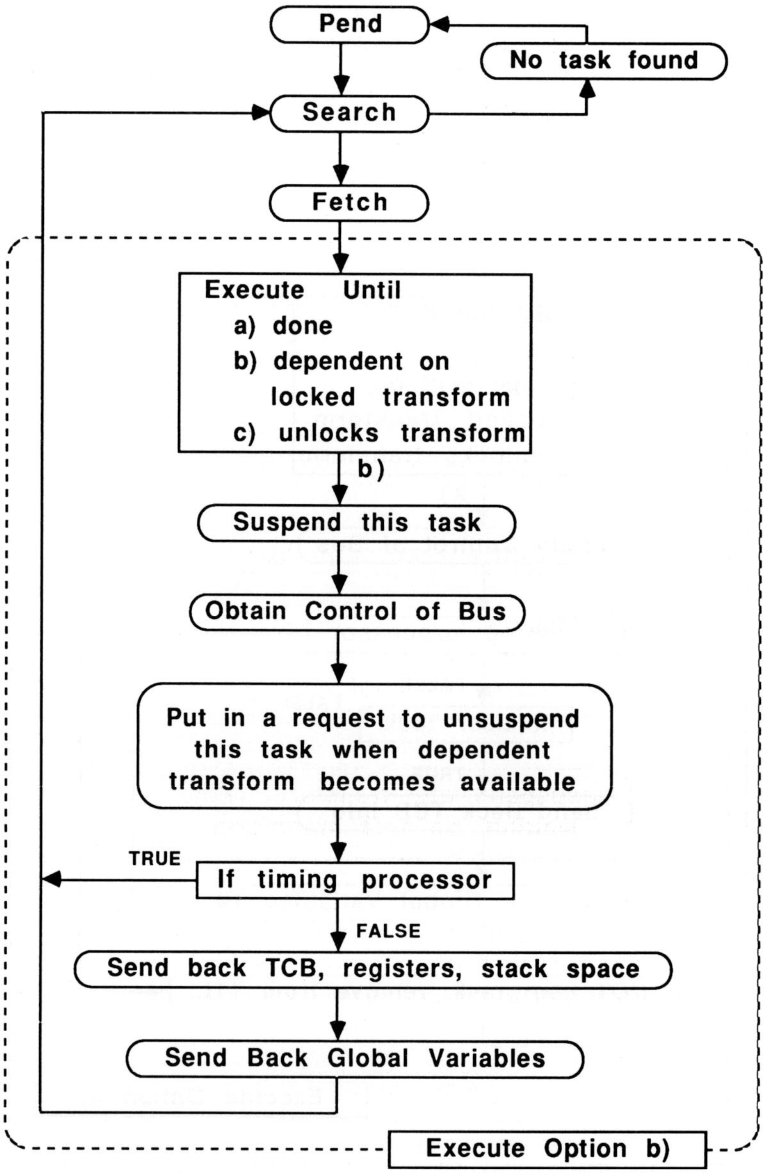

Figure 17b. The b) option of the execute state of the set-point algorithm.

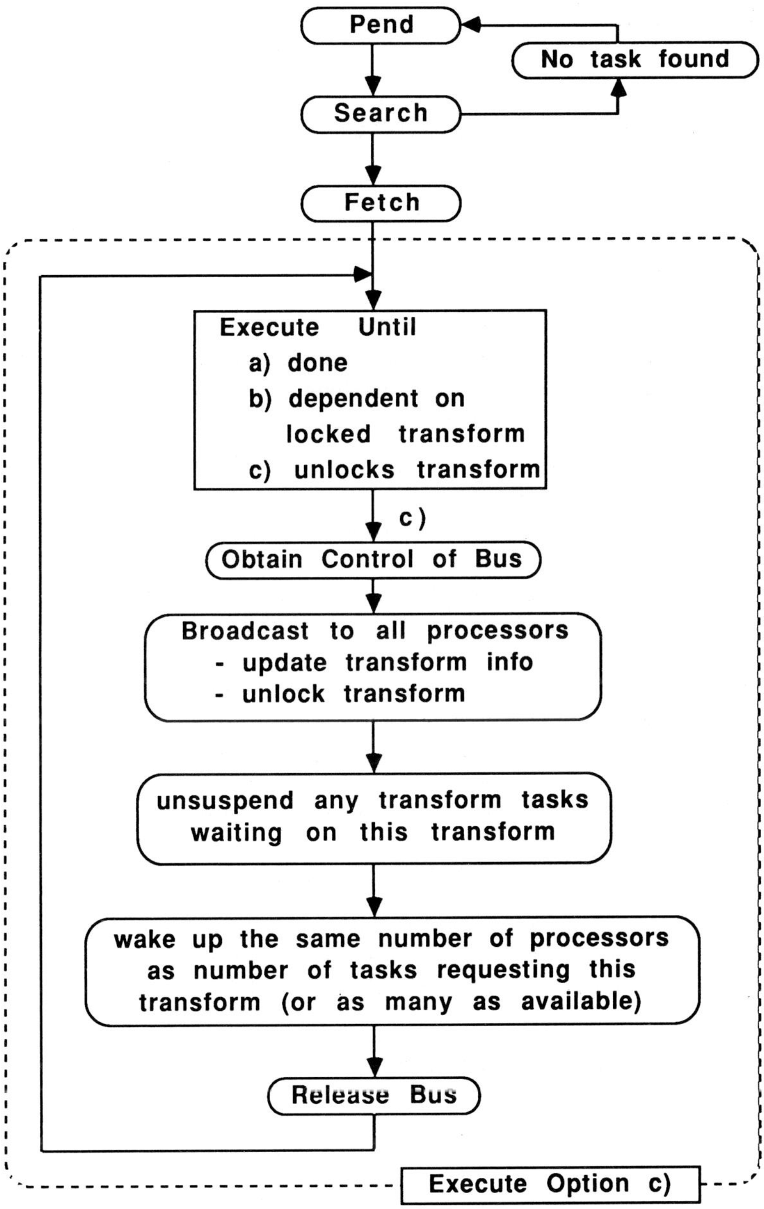

Figure 17c. The c) option of the execute state of the set-point algorithm.

Step 6: Execute the transform task until one of three things happen:

a) the task is completed, in which case obtain access to the bus, remove the transform task from the TTL and go to step 2.

b) the task execution depends on a locked transform and can not continue any further, in which case suspend the current task, obtain control of the bus and put a request in to unsuspend this task when the dependent transform becomes unlocked. Then send back the Task Control Block information and the global variables, then go to step 2.

c) the task has computed the elements of the transform, in which case obtain control of the bus, then write the updated transform information and unlock the copy of the transform in every processor's shared RAM. Then unsuspend any transform tasks which put in a request for this transform, and wake up the same number of processors as number of tasks requesting this transform (or as many processors as available), then release the bus and continue with step 6.

8.6 The Trajectory Algorithms Revisited

In the previous sections, we derived the algorithms used to provide coordinating motion for multiple robotic devices. We divided the motion into motion segments, each consisting of a primary and a secondary motion. Given that these two motions are computed independently of each other and then added to achieve the final motion, we can split the computation among separate processors. The following is a summary of the equations of motion and how they are split between the primary trajectory generator routine and the secondary trajectory generator routine.

The four transition cases are summarized below as divided between the **primary** and **secondary** trajectory generator routines.

Joint motion to joint motion

primary: $\vec{\theta}_{out} = r(\vec{\theta}_{BC}) + \text{device}{\rightarrow}K^{-1}(\Gamma(P_C(t)), \text{configuration})$ (101a)

secondary: $+ ((((a_7 p + a_6)\, p + a_5)\, p + a_4)\, p^3 + a_1)\, p + a_0$ (101b)

where the coefficients of the polynominal are determined in joint space.

Joint motion to Cartesian motion

primary: $\vec{\theta}_{out} = \text{device}{\rightarrow}K^{-1}\ (\Gamma(L_C(t)\ D(r)\ R_C(t)),\ \text{configuration})$ (102a)

secondary: $+\ ((((a_7\ p + a_6)\ p + a_5)\ p + a_4)\ p^3 + a_1)\ p + a_0$ (102b)

Cartesian motion to Cartesian motion

primary: $T_n = \Gamma(L_C(t)\ D(r)\ R_C(t)),\ \text{configuration})$ (103a)

secondary: $T_{matching} = \Gamma(L_T\ D_s(p)\ R_T),\ \text{configuration})$ (103b)

primary: $T_n \mathrel{*}= [T_{matching}]^{-1} [T_B(-t_{acc})]$ (103c)

primary: $\vec{\theta}_{out} = \text{device}{\rightarrow}K^{-1}\ (T_n,\ \text{configuration})$ (103d)

Cartesian motion to joint motion

primary: $\vec{\theta}_{out} = r(\vec{\theta}_{BC}) + \text{device}{\rightarrow}K^{-1}\ (\Gamma(P_C(t)),\ \text{configuration})$ (103e)

secondary: $+\ ((((a_7\ p + a_6)\ p + a_5)\ p + a_4)\ p^3 + a_1)\ p + a_0$ (103f)

When the TTL is built at the end of each sample time, the primary trajectory generator is associated with the device transform, and during transitions the secondary trajectory generator is associated with a different transition task transform which can be executed in parallel by another set-point processor.

9.0 Processor System Considerations

By dividing the trajectory generator routines into primary and secondary routines, we also divide the amount of global variables associated with each device. In each case 2K bytes of memory is sufficient to store all the global information needed. If we let 2K bytes be the average size of the global variables needed for all transform tasks, we can roughly compute the size of shared RAM required. The variables associated with each transform task, shown in Figure 15, including the Task Control Block, the floating point registers, and the function stackspace add up to 2K bytes in size. If we let the maximum number of robotic devices we would want to control with one controller be four, and let each device have four active transforms each sample period, then this adds to 40K bytes for the transform task space, remembering to count the devices one time each as they will have transforms on the TTL.

If we allow 1K bytes for system variables such as mailboxes and system addresses and 8K bytes for global variables for the robot devices, this adds to 49K bytes. If we use a 64K bytes of shared RAM, this leaves 15K for user global variables. If this is not enough, more shared RAM should be used.

Given a Multibus II or VME system bus, we can reasonably transfer data at 10 megabytes per second across the bus. This translates into 10K bytes in 1 millisecond. From the discussion above, the transform task adds up to 2K bytes in size, and the global variable size is 2K on the average. Therefore, we will use 0.5 milliseconds for a bus transfer time allowing 0.1 milliseconds for DMA overhead. If more than 2K bytes of memory for global variables are needed for some transform tasks, or if the execution time of the *defining_function* is very small, one option would be to restrict those tasks to run on the same processor, with no bus transfer.

The processor with the highest priority is chosen as the timing processor. Its function is identical to the other processors except its timing function. After the executable code is downloaded to the multiprocessor system, the timing function starts the real-time clock which defines the sample time for the system. When the timer defining the sample time counts down to zero, it interrupts the timing processor, which is pending on this event. The timing processor signals every other processor on the bus that a new set-point period has begun. From this point on, every set-point processor executes in an identical manner during the current set-point.

When the executable file is downloaded from the host, the head of the Transform Task List is configured as a standard memory location in the shared RAM and this address is written into the shared RAM of every every processor. Before the start of execution, the TTL only contains a transform for each user program downloaded. Initially no other transforms exist because the user program initializes the controllable devices and creates the transforms needed for a given robot task.

10.0 Performance Evaluation

To examine the performance of the controller, we will look at the execution characteristics of a single six axes manipulator and also an application involving coordinated control of three robotic devices in an inspection scanning task explained in section 4.1.

We will start with the TTL which the controller would produce for the task, and then artificially break the Transform Tasks into sub-tasks where the dependences occur. This enables us to roughly compare the performance of our task partitioning method with the DF/IHS method devised by Kasahara and Narita (Kasahara & Narita 85). It is useful to keep in mind the differences between the set-point algorithm and the DF/IHS algorithm when looking at the results. The set-point algorithm knows when a task has been suspended, and therefore when it is necessary to transfer the TCB and related information. The DF/IHS always transfers this information. The set-point algorithm knows that a subtask belongs to an already executing transform task on the same processor and there is no need to transfer new global variables as the same global variables are used for the entire transform task. The DF/IHS method takes each subtask as independent in terms of the global variables it needs and loads them with each new subtask. This is often the case for the DF/IHS method because it rarely schedules consecutive subtasks on the same processor. Finally, in deriving the DF/IHS method, the authors assumed identical processors and a common memory not associated with any one processor. Therefore the fetch and return times are independent of the processor. For purposes of comparison, we will run our method first with this assumption, and then with the assumption the shared memory is on the timing processor which has zero fetch and return times.

The biggest difference between the set-point algorithm and the DF/IHS algorithm is that the set-point algorithm schedules the tasks in real-time. The DF/IHS method must run on a main frame and is not necessarily suitable for general real-time scheduling (Kasahara & Narita 85).

In the kinematic control of a single six axes manipulator, the PUMA for example, two extremes of computation requirements exist. The least demand is during the linear portion of a Cartesian move, equation (102a). For this the Transform Task List appears in Table 8. The execution times are taken from section 5.0. The execution times for two processors with the above TTL is shown in Figure 18. We examine this case to show that the trajectory generation takes place on one processor allowing the second processor to run one or multiple *user_tasks()*. It may be possible to use this extra computation power to run a sensor coordinator program.

Table 8. Transform Task List for Linear region, PUMA robot.

TTL	Task Number	Predecessor	Name	Time (ms)
Puma	0	none	dummy	0.0
	1	0	D(r)	0.4
	2	1	Γ	2.0
	3	2	K^{-1}	2.0
	4	3	misc	1.0
User Task	5	0	user	2.0
	6	4, 5	exit	0.0

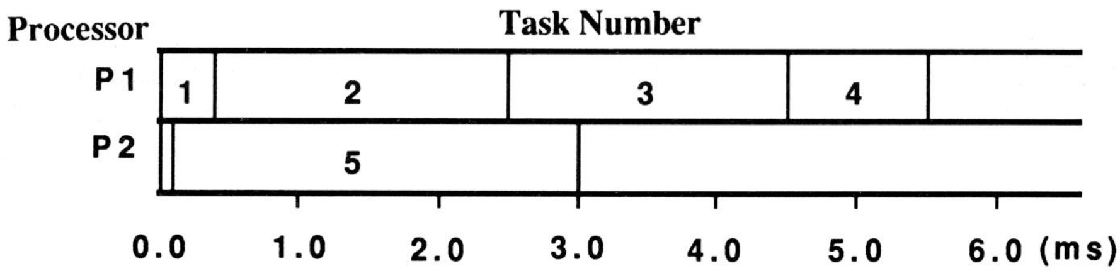

Figure 18. The execution profile using two processors for the tasks shown in Table 8.

If we use a uniform sample time, the worst case sample period determines the task sample time. The worst case in terms of computational load occurs during the transition from Cartesian mode to Cartesian mode equations (103a,b,c,d). During that period the Transition Task List can be represented as shown in Table 9.

The task dependency graph is shown in Figure 19. Execution times for one through five processors is illustrated in Figure 20. The three lines on the graph correspond to the DF/IHS method, the set-point algorithm with common memory not on any set-point processor, and the set-point algorithm with shared memory on the timing processor. If we do not run the user task during this one sample period, a uniform sample time of eight milliseconds can be achieved using two processors.

In the boat hull scanning example in section 4.1, we coordinate three devices, a one axis table, and two five axes manipulators. Using the position equations 82, 83, and 84, we can form the TTL for the scanning region with five Transform Tasks, shown in Table 10.

The task dependency graph for the scanning system is shown in Figure 20. From the graph shown in Figure 21, we see that the optimal number of processors for this scanning task is three. After that the small gain in sample time is offset by the cost of more hardware.

Table 9. Transform Task List for the cartesian to cartesian transition for the PUMA robot.

TTL	Task Number	Predecessor	Name	Time (ms)
Puma	0	none	dummy	0.0
	1	0	D(r)	0.4
	2	1	Γ	2.0
	3	2,6	K^{-1}	2.0
	4	3	Δ	1.8
	5	4	misc	1.0
Transition	6	0	Γ_T	0.5
	7	6	Δ_s	4.5
User Task	8	0	user	2.0
	9	5, 7, 8	exit	0.0

Table 10. Transform Task List for Scanning System.

TTL	Task Number	Predecessor	Name	Time (ms)
Table	0	none	dummy	0.0
	1	0	Γ_{start}	0.5
	2	1, 4	Γ_{end}	1.0
	3	2	K^{-1}	0.6
Scan Trans	4	0	orient	2.0
Master	5	0	Γ_{start}	0.5
	6	2, 5	Γ_{end}	1.5
	7	6	K^{-1}	2.0
Slave	8	0	Γ_{start}	0.5
	9	6, 8	Γ_{end}	1.5
	10	9	K^{-1}	2.0
User Task	11	0	user	2.0
	12	3, 7, 10, 11	exit	0.0

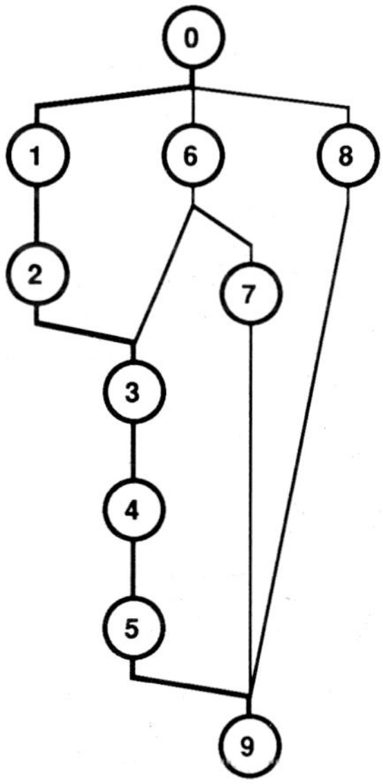

Figure 19. The task dependency graph for the tasks shown in Table 9.

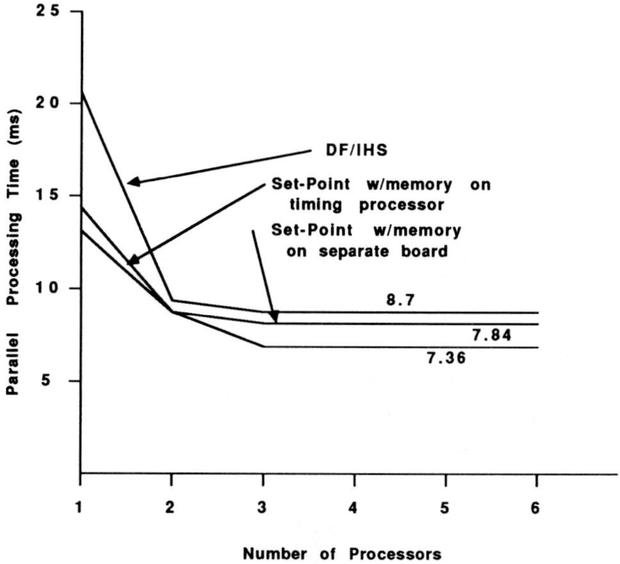

Figure 20. The execution profile for the tasks shown in Table 9.

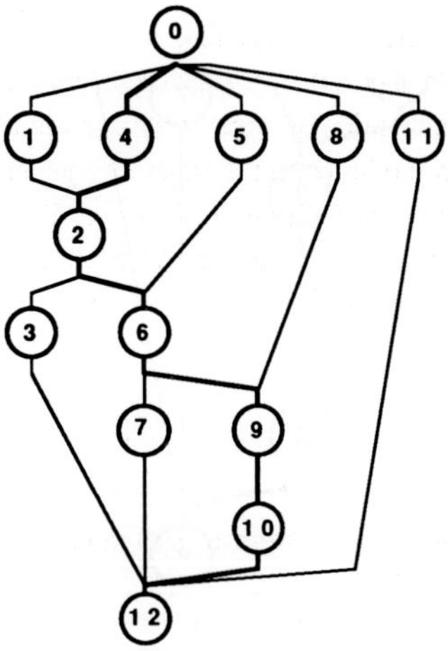

Figure 21. The task dependency graph for the tasks shown in Table 10.

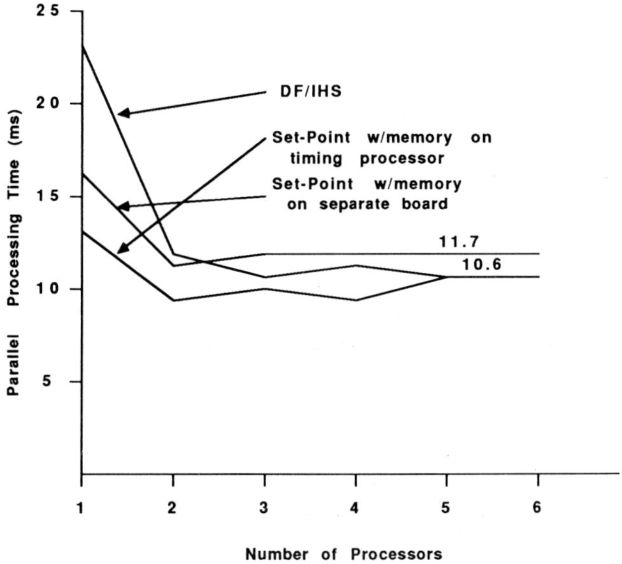

Figure 22. The execution times for the tasks shown in Table 10.

11.0 Chapter Summary

In this chapter we have outlined the mathematical basis for coordinated kinematic control of multiple robotic devices. We have also discussed a possible pseudo code implementation consisting of a device independent trajectory generator and the ability to time-slice user-written tasks. The position specification method and the trajectory generator allow motion coordination while the time-sliced user tasks allow communication coordination and programming flexibility. All robotic devices which are available for motion coordination are kept on a system device list. During actual execution of a robot task, those devices needed for the task are placed on the device task list.

Executing these equations of motion on our multiprocessor system gives us two advantages over a single processor system. First, it yields a higher sample rate which translates into faster sensor response time for the robotic devices. Second, our multiprocessor system can act as a redundant controller, allowing the system to still function, albeit at a lower sample rate, should a processor other than the timing processor fail. By recognizing that during the execution of Cartesian task each sample period involves the evaluation of the transforms currently specifying motion, we created a Transform Task List (TTL) which totally specifies these tasks. The set-point algorithm developed in this chapter schedules these tasks in real-time allowing each processor on the tightly-coupled architecture to executes these tasks in an efficient manner. This method of task allocation allows robot tasks of varying computational requirements to run efficiently on a system with a user selectable number of processors.

The execution times of tasks scheduled in real-time with the set-point algorithm are similar to those obtained off-line with the DF/IHS algorithm. The set-point algorithm achieves its performance by using knowledge of the task and by having the shared RAM located on the timing processor board.

The sample time we can achieve depends on the robot task and the number of processors used. For purposes of comparison, we have shown an eight millisecond sample time is achievable with two set-point processors with ample time left to run one or several users tasks as well as computationally intensive functionally defined transforms.

LIST OF REFERENCES

Ahmad, S., Li, B., (1989), "Robot Control Computation in Microporcessor Systems with Multiple Arithmetic Processors using a Modified DF/IHS Algorithm," IEEE Transactions on Systems, Man, Cybernetics, Vol. 19, no. 5, September 1989, pp. 1167-1178.

Ahmad, S., and Besant, C., (1984), "Robot Level Programming Languages and the SRIL-90 Language," Proceedings of the IEEE Compsac Conference, Nov. 1984, Chicago, IL.

Ahmad, S., and Guo, H., (1989), "Dynamic Coordination of Dual-Arm Robotic Systems with Joint Flexibility," Proceedings of the IEEE Conference on Robotics and Automation, April 1989, Philadelphia, PA.

Ahmad, S., and Luo, S., (1989), "Coordinated Motion Control of Multiple Robotic Devices for Welding and Redundancy Coordination through constrained Minimization in Cartesian Space," IEEE Transactions on Robotics and Automation, Vol. 5, No. 4, Aug. 1989, pp. 409-417.

Bejczy, A. K., (1974). "Robot arm dynamics and control," tech. memo 33-669, Jet Propulsion Laboratory.

Gajski, D. D., and Peir, J., (1985), "Essential Issues in Multiprocessor Systems," IEEE Computer, Vol. 18, No. 6, pp. 9-27.

Goldenberg, A. A., Benhabib, B., and Fenton, R. G., (1985), "A Complete Generalized Solution to the Inverse Kinematics of Robots," IEEE Journal of Robotics and Automation, Vol. RA-1, no. 1, pp. 14-20.

Gupta, K. C., (1986), Kinematic Analysis of Manipulators Using the Zero Reference Position Description," The International Journal of Robotics Research, Vol. 5, No. 2.

Guptill, R. L., and Ahmad, S., (1990), "Programming Path Specific Robot Operations with Optimal Part Placement," TR-EE-90-24, Purdue University, West Lafayette, Indiana, USA.

Guptill, R. L., and Stahura, P. A., (1987), "Multiple Robotic Devices: Position Specification and Coordination", IEEE Conference on Robotics and Automation, North Carolina.

Hayati, S., (1986), "Hybrid Position / Force Control of Multi-Arm Cooperating Robots," IEEE Conf. on Robotics and Automation, San Francisco.

Hayward, V., (1983), "RCCL User's Manual, Version 1.0", TR-EE 83-46, School of Electrical Engineering, Purdue University.

Hayward, V. and Hyati, S., (1988), "Kali: An Environment for Programming and Control of Cooperative Manipulators," CIM-88-8, Technical Report of the McGill Research Center for Intelligent Machines, McGill University, Montreal, Canada.

Hayward, V., and Paul, R., (1984), "Introduction to RCCL: A Robot Control "C" Library", Proc. IEEE Conf. on Robotics. pp. 293-297, Atlanta.

Intel, DEC, and Xerox, (1982), "Ethernet Data Link and Physical Layer Specification, Version 2.0," Intel, Dec, and Xerox.

Kasahara, H., and Narita, S., (1985), "Parallel Processing of a Robot Arm Control Computation on a Multiprocessor System," IEEE Journal of Robotics and Automation, RA-1, Number 2.

Liu, C-H., and Chen, Y-M., (1986), "Multi-Microprocessor-Based Cartesian-Space Control Techniques for a Mechanical Manipulator," IEEE Journal of Robotics and Automation, Vol. RA-2, No. 2, pp. 110-115.

Luh, J. Y. S., Walker M. W., and Paul, R., (1980), "On-line computational scheme for mechanical manipulators," Trans. ASME, Journal of Dynamic Systems, Measurement and Control, vol. 120, pp. 69-76.

Luh, J. Y. S., and Lin, C. S., (1982), "Scheduling of Parallel Computation of Computer Controlled Mechanical Manipulator," IEEE Transactions on Systems, Man, and Cybernetics, SMC-12, 214.

McCarthy, J. M., (1986), "Dual Orthogonal Matrices in Manipulator Kinematics," The International Journal of Robotics Research, Vol. 5, No. 2, pp. 45-51.

Nigam, R., and Lee, C. S. G., (1985), "A Multiprocessor-Based Controller for the Control of Mechanical Manipulators", IEEE Journal of Robotics and Automation, RA-1, Number 4, pp. 173-182.

Paul, R. P. (1981), Robot Manipulators: Mathematics, Programming, and Control , Cambridge: MIT Press.

Paul, R. P., and Zhang, H., (1985), "Robot Motion Trajectory Specification and Generation," Robotics Research: The Second International Symposium, MIT-Press Series in Artificial Intelligence.

Pieper, D. L., (1968), "The Kinematics of Manipulators Under Computer Control," Stanford Artificial Intelligence Laboratory, Stanford University, AIM 72.

Taylor, R. H., (1979), "Planning and Execution of Straight-Line Manipulator Trajectories," IBM J. Research and Development, Vol. 23, pp. 424-436.

Zhang, H., (1986), "Design and Implementation of a Robot Force and Motion Server", PhD. Thesis, Purdue University.

CHAPTER 6

Neural Networks for Real-Time Robotic Applications

Nikzad Benny Toomarian
Jacob Barhen‡
Sandeep Gulati

Jet Propulsion Laboratory
California Institute of Technology
Pasadena, CA 91109

ABSTRACT

A fundamental problem in computational AI – developing a new class of massively parallel, neural learning algorithms for robustly abstracting complex nonlinear transformations from representative samples, in real-time, is addressed. Provision of such a capability is at the core of many difficult problems in robotics, signal processing, remote sensing, control, etc. Our neural learning formalism builds upon the concept of adjoint operators, to enable a fast computation of the gradients of an energy functional with respect to all parameters of the neural architecture, and exploits the concept of *teacher forcing* to incorporate information regarding the desired output into the activation dynamics. The powerful repertoire of adaptive neural learning formalisms presented in this paper are then exploited to provide an enabling core for addressing a fundamental problem in robotics - the design of autonomous robots designed to perform tasks in unstructured and unpredictable environments. Furthermore, proposed methodology provides the opportunity for future hardware implementation of real-time robotic applications.

‡ Also with the Applied Physics Department, Calif. Inst. of Tech.

1. INTRODUCTION

In the past few years the quest for efficient computational approaches to artificial intelligence and cognitive engineering has undergone a significant evolution. Not only is this transformation, from discrete symbolic reasoning to massively parallel connectionist neuroprocessing, of compelling scientific interest, but it also is of tremendous practical interest. It is changing the very nature of information processing and problem solving. In general, the scientific and engineering community is confronted with two basic categories of problems. First, there are problems that are clearly defined and deterministic. This class deals with *structured problems* such as sorting, data processing and automated assembly in controlled workspace. On the other hand, there are scenarios such as maintenance of nuclear plants, undersea mining, battle management and assembly/repair of space satellites, that lead to computational problems that are inherently ill-posed and ill-conditioned. Such *unstructured problems* entail providing for situations that may have received no prior treatment or thought. In summary, the latter category encapsulates problems that cannot be satisfactorily addressed using traditional computational paradigms such as Random Access Machines, Markov Algorithms, Universal Turing Machines, Cellular Automata, Recursive Function Theory or Production Systems. The main motivation for this work is by the current lack of computationally enabling tools for engineering "intelligent autonomous systems" - autonomous robots, rovers, diagnostic systems etc. Robotics in general, and robust, task-directed, autonomous manipulation for space-based robots in particular, represent our specific application domains of interest.

Engineered intelligent systems, e.g., autonomous robots and rovers for space applications, behave with remarkable rigidity when compared to their biological counterparts, especially in their ability to recognize objects or speech, to manipulate and adapt in an unstructured environment and to learn from past experience. They lack common sense knowledge and reasoning, and knowledge structures for recognizing complex patterns, i.e., they fail to recognize their own limitations. On the other hand, the biological machinery is capable of providing satisfactory solutions to such hard problems with remarkable ease and flexibility [3,16,17]. A key emphasis underlying any paradigmatic development for unstructured computation today, is to understand how the aforementioned unstructured computations are interpreted, organized and carried out by the biological systems. The latter exhibit a spontaneous emergent ability that enables them to self-organize and adapt their structure and function.

It is rapidly becoming evident that many unstructured problems, can best bet solved by "analogy", "subsymbolic" [39] or pattern matching techniques [46]. While artificial intelligence attempts to do this, *neural networks*, a biologically inspired, computational and information processing paradigm, provides elements of with an inherently better, but not a unique tool. In the remainder of this work, we concentrate on developing a repertoire of computational formalisms that can provide us an enabling basis for solving a significantly complex problem, that has been addressed for last several decades, namely, *functional synthesis, i.e., learning nonlinear mappings* with the goal to abstract functional, statistical, logical and spatial invariants from representative examples.

The remainder of this chapter work is organized along the following lines. In Section 2, we briefly characterize the phenomenology of artificial neural networks. The teacher forcing paradigm which enables the network to be trained in a fast manner is presented in Section 3. In Sections 4 and 5, we derive two versions, temporal and static, of supervised learning algorithms for dynamical neural networks. Section 6 reports on the application of the static algorithm to simulate the 3-DOF and 7-DOF redundant manipulators. We conclude the work by presenting some final thoughts and remarks in section 7.

2. ARTIFICAL NEURAL NETWORKS

Recent advances in our understanding of the anatomical and functional architecture [4,5] of the electrical and organizational processes [25-26] occurring in the brain and nervous system, along with advances in hardware technology and capabilities are leading to physical and electro-optical realizations [20] of artificial neural networks. These networks posses computationally useful properties such as time sequence retention, error correction, noise elimination, recognition and generalization. The resurgence of the field is due to the more recent theoretical contributions by Amari [2], Carpenter [12], Grossberg [18], Hopfield [23], Kohonen [30],etc. Hopfield's illuminating contributions have extended the applicability of neuromorphic techniques to the solution of combinatorially complex optimization problems [24]. In the areas of VLSI and opto-electronic implementations, major achievements have resulted from the efforts of Mead [36], Psaltis and Farhat [38], Yariv, Agranat and Neugebauer[48], and others.

As shown in Fig. 2.1, an *Artificial Neural System* may be characterized as a distributed computational system comprising a large number of processing units, each of which has selected characteristics of biological neurons connected to each other in a directed graph of varying configuration [28]. They have been defined [30] as *"massively parallel, adaptive dynamical systems modeled on the general features of biological networks, that can carry out useful information processing by means of their state response to initial or continuous input. Such neural systems interact with the objects of the real world and its statistical characteristics in the same way as biological systems do".* Neural networks have also been characterized as systems that are (a) nonlinear, (b) nonlocal, i.e., exhibit long-range interactions across a network of locations, (c) nonstationary, i.e., interactions are reverabative or iterative, and (d) modeled in terms of a nonconvex, "energy-like", function [18].

The advantages of neuronal processing arise as a result of their ability to perform massively parallel, asynchronous and distributed information processing. Neurons with simple properties and interacting according to relatively simple rules can accomplish collectively complex functions. This is based on their ability to provide a collectively-computed solution to a problem on the basis of analog input information resulting from a high degree of random interconnectivity, storage and simplicity of individual operations [8]. Neural network modeling then is a discipline which attempts to "understand brain- like computational systems" and has been variously termed as "computational neuroscience", "parallel distributed processing" or "connectionism".

The bulk of neural network models can be classified into two categories; those that are intended as computational models of biological nervous systems or *neurobiological models*, and those that are intended as biologically-inspired models of computational devices with technological applications, also referred to as *Artificial Neural Systems (ANS)* . In this work, our attention is focused on ANS.

The above phenomenological insights at the neurobiological level, have led to the mathematical formulation of *simulated neurons*, i.e., the basic building block of neural network models. A functional model for a typical simulated neuron is shown in Fig. 2.2. Four useful areas may be abstracted. The first is the synapse where signals are passed from one neuron to another, and the amount of the signal is regulated, i.e., gated or weighted by the strength of the synaptic interconnection. In the activated neuron region, denoted as *summer*, synaptic signals containing excitatory and inhibitory information are combined to affect the tendency of a neuron to fire or stay put. The threshold detector determines if the neuron is actually going to fire or not, while axonal paths conduct the output activation energy to other synapses to which the neuron is connected. Useful information properties such as generalization [22], classification [11], association [1,14], error correction, time sequence retention [34], etc., emerge as collective properties of systems comprised of aggregations of large number of such simple units. When viewed individually, the dynamics of each neuron bears little semblance to task being performed.

As discussed in the preceding paragraph, of particular computational and modeling interest, are the mathematical notions of synapse and synaptic modification, mechanisms by which such units can be connected together to compute, and the rules whereby such interconnected systems could be made to learn.

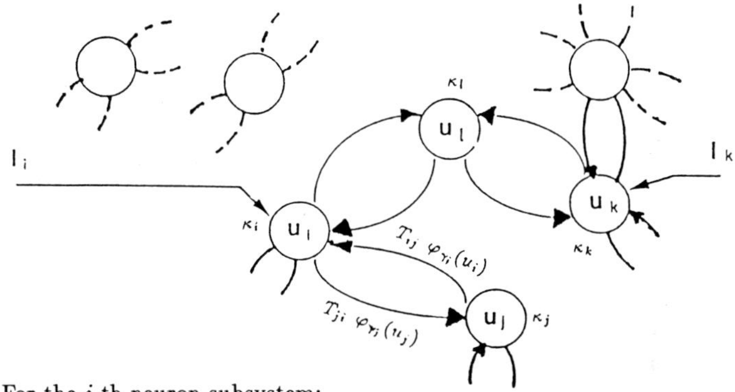

For the i-th neuron subsystem:

u_i : i-th neuron's "potential" (activity level)

T_{ij} : "synaptic weight" of the connection from neuron j to neuron i

$\varphi_{\gamma i}$: "firing rate" function

γ_i : Neural gain

I_i : External input

κ_i : Internal decay constant

Figure 2.1 Artificial Neural System.

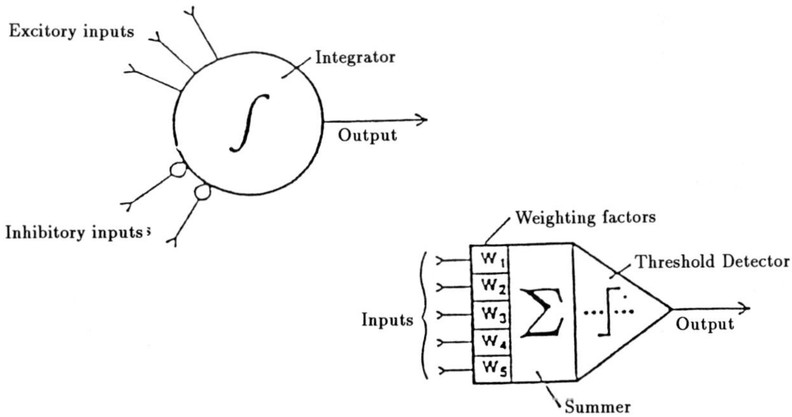

Figure 2.2 Mathematical idealization to biological neurons.

2.1. Computational Learning

The paradigmatic strength of neural networks for potential applications, which require adaptive solution of intractable computational or modeling problems, arises from their spontaneous emergent ability to achieve *functional synthesis*, and thereby learn nonlinear mappings [7], and abstract the spatial [12], functional [32] or temporal [31,43,44] invariances of these mappings. Thus, relationships between multiple continuous-valued, statistically-related inputs and outputs can be established, based on a presentation of a large number of representative examples. Once the underlying invariances have been learned and encoded in the topology and strengths of the synaptic interconnections [7], the neural network can generalize to solve arbitrary problem instances. Since the topological mappings for problem-solving are acquired from real-world examples, network functionality is not limited by assumptions regarding parametric or environmental uncertainty, that invariably limit model-based computational strategies [2]. In order to place our subsequent discussions in context with the overall discipline of computational learning we provide a taxonomy of learning formalisms.

Neural Learning has been defined as the process of adaptively evolving the internal parameters e.g., connection weights, network topology, etc., in response to stimuli being presented at the input and, possibly, the output buffer. As adaptive dynamical systems, neural networks emphasize relaxation, and not heuristic search, as the basis of automatic learning. Further, learning can be achieved either by programming the network, i.e., setting the weights, or by self-organization. These are known as *hard* and *soft* learning respectively. In this work we focus on soft learning, which in turn may be either deterministic or stochastic, and either supervised [7] or unsupervised [45,12]. Learning in neural networks is said to be *supervised*, when the desired response is from a knowledgeable teacher, and the retrieval involves one or more of a set of stimuli patterns that have been repeatedly shown to the system during the training phase.

2.2 Neural Learning Framework

We formalize a neural network as an adaptive dynamical system whose temporal evolution is governed by the following set of coupled nonlinear differential equations:

$$\dot{u}_n + \kappa_n u_n = g_n[\gamma_n(\sum_m T_{nm} u_m + I_n)] \qquad t > 0 \tag{2.1}$$

where u_n represents the output of the nth neuron ($u_n(0)$ being the initial state), and T_{nm} denotes the strength of the synaptic coupling from the $m-$th to the $n-$th neuron. The constants κ_n characterize the decay of neuron activities. The sigmoidal functions $g_n(\cdot)$

modulate the neural responses, with gain given by γ_n; typically, $g_n(\gamma_n x) = \tanh(\gamma_n x)$. In order to implement a nonlinear functional mapping from an N_I-dimensional input space to an N_O-dimensional output space, the neural network is topographically partitioned into three mutually exclusive regions. As shown in Figure 2.3, the partition refers to a set of input neurons S_I, a set of output neurons S_O, and a set of "hidden" neurons S_H. Note that this architecture is not formulated in terms of "layers", and that each neuron may be connected to all others including itself. To specify the term I_n we distinguish between two different cases:

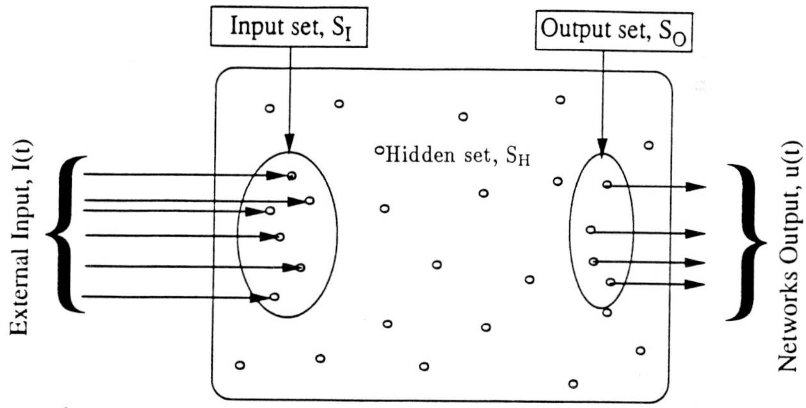

Fig. 2.3 - Topographic Partition of a Neural Network

1.- A temporal patern is presented to the network, in which case, $\bar{a}(t)$ (henceforth the overhead bar will denote a vector) is an N-dimensional vector of target temporal patterns, with non zero elements, $a_n(t)$, in the input and output sets only. When trajectories, rather than mappings, are considered, as we shall see in the sequel, components in the input set may also vanish. Note that, the time-dependent external input term in Eq. (1), i.e., $I_n(t)$, encodes component-contribution of the target temporal pattern in the input set only, i.e.,

$$I_n(t) = \begin{cases} a_n(t) & \textbf{if } n \in S_I \\ 0 & \textbf{if } n \in S_H \cup S_O \end{cases} \tag{2.2}$$

To proceed formally with the development of a temporal learning algorithm, we consider an approach based upon the minimization of an error functional, E, defined over the time interval $[t_o, t_f]$ by the following expression

$$E(\bar{u}, \bar{p}) = \int_{t_o}^{t_f} \frac{1}{2} \sum_n e_n^2 \, dt = \int_{t_o}^{t_f} F dt \tag{2.3}$$

where the error component, $e_n(t)$, represents the difference between the desired and actual value of the output neurons, i.e.,

$$e_n(t) = \begin{cases} a_n(t) - u_n(t) & \textbf{if } n \in S_O \\ 0 & \textbf{if } n \in S_I \cup S_H \end{cases} \tag{2.4}$$

2.- Static paterns are presented to the network, in which case, the "source" term ${}^kI_n$, encodes contribution of the k-th training sample via the following expression

$$ {}^kI_n = \begin{cases} {}^ka_n & \textbf{if } n \in S_I \\ 0 & \textbf{if } n \in S_H \cup S_O \end{cases} . \tag{2.5}$$

As in the temporal case, the development of a static supervised learning algorithm, requires minimization of a error function E given by the following expression

$$E(\bar{u}, \bar{p}) = \frac{1}{2} \sum_k \sum_n [\, {}^k\tilde{u}_n - {}^ka_n \,]^2 \qquad \forall\, n \in S_I \cup S_O \tag{2.6}$$

Where vector $\tilde{u}$ denots the steady state solution of the neural activation dynamcis, Eq. (2.1).

In both static and temporal models, the internal dynamical parameters of interest are the strengths of the synaptic interconnections, T_{nm}, the characteristic decay constants, κ_n, and the gain parameters, γ_n. They can be represented as a vector of M $[M = N^2 + 2N]$ components

$$\bar{p} = \{\, T_{11}, \cdots, T_{NN}, \; \kappa_1, \cdots, \kappa_N, \; \gamma_1, \cdots, \gamma_N \} \tag{2.7}$$

We will assume that elements of $\bar{p}$ are statistically independent. Furthermore, we will also assume that, for a specific choice of parameters and set of initial conditions, a unique solution of Eq. (2.1) exists. Hence, the state variables $\bar{u}$ are an implicit function of the parameters $\bar{p}$. In the rest of this paper, we will denote the μth element of the vector $\bar{p}$ by p_μ $(\mu = 1, \cdots, M)$.

3. TEACHER FORCING

The neural activation dynamics specified by Eqs. (2.1), (2.2) and (2.5) do not include explicit information regarding the desired network states. If these equations are used in conjunction with the learning formalism described in the sequel, the network parameters (i.e., the elements of $\bar{p}$) will be modified either at the end of a trajectory as in the temporal case, or after all the patterns have been presented to the network in the static case. This is illustrated schematically in Figure 3.1. Network parameter adaptation is based upon the total error between the desired and the actual state of the network. Note that conventional supervized learning algorithms operate in this fashion, and usually require excessive iterations to produce the desired results. In order to overcome this difficulty, we consider the concept of *teacher forcing*, i.e., driving the neurons to desired values in finite time. To the best of our knowledge, the work of Williams and Zipser [47] is the only instance where forcing has been used in a similar context.

The teacher forcing paradigm proposed here, stems from feedback control. In such a scheme, with continuous network dynamics, the error between the actual and the desired neurons values is fed back, as input to the network. A schematic block diagram of our model, for the temporal case, is presented in figure 3.2. The parameters of the network are updated based upon the error accumulated over the length of the trajectory, i.e., over the interval $[t_o, t_f]$. In order to incorporate this teacher forcing into the neural learning formalism, the time-dependent input to the neural activation dynamics is modified to read:

$$I_n(t) = \begin{cases} a_n(t) & \textbf{if } n \in S_I \\ 0 & \textbf{if } n \in S_H \\ \lambda[a_n(t)]^{1-\beta}[a_n(t) - u_n(t)]^{\beta} & \textbf{if } n \in S_O \end{cases} \tag{3.1}$$

At this stage, λ and β are assumed to be positive constants. The purpose of the term $[a_n(t)]^{1-\beta}$ is to insure that $I_n(t)$ has the same dimension as $a_n(t)$ and $u_n(t)$.

In the static case, however, we have traditionally considered teacher forcing induced from the input set, S_I, rather than the output set, S_O. This is not essential, and, as for temporal learning, the converse can be readily considered. Thus, the "source" term ${}^kI_n$, is modefied to encode contribution in terms of attractor coordinates of the k-th training sample via the following expression

$${}^kI_n = \begin{cases} \lambda[{}^ka_n]^{1-\beta}\,[{}^ka_n - {}^k\tilde{u}_n]^{\beta} & \textbf{if } n \in S_I \\ 0 & \textbf{if } n \in S_H \cup S_O \end{cases} \tag{3.2}$$

Zak[49] has demonstrated that in general, for $\beta = (2i+1)/(2j+1)$, $i < j$ and i and j strictly positive integers, an expression of the form $[a_n - u_n]^{\beta}$ induces a terminal attractor phenomenon for the dynamics described be Eq. (2.1).

When teacher forcing is implemented in S_O, and learning is successfully completed, teacher forcing will vanish, and the network will revert to the conventional dynamics

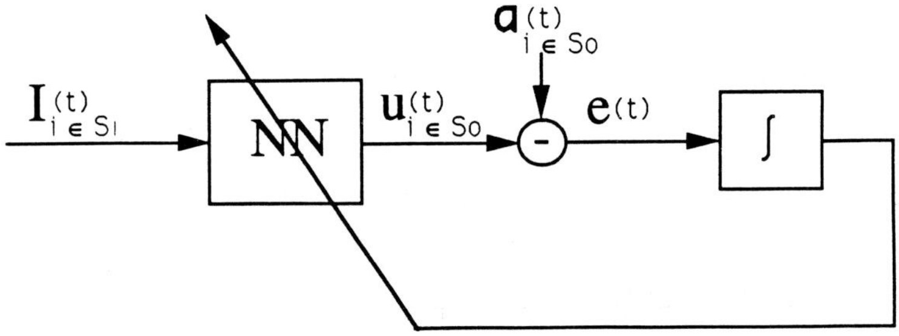

Fig. 3.1 - Schematic Diagram of Learning Algorithm Without Teacher Forcing

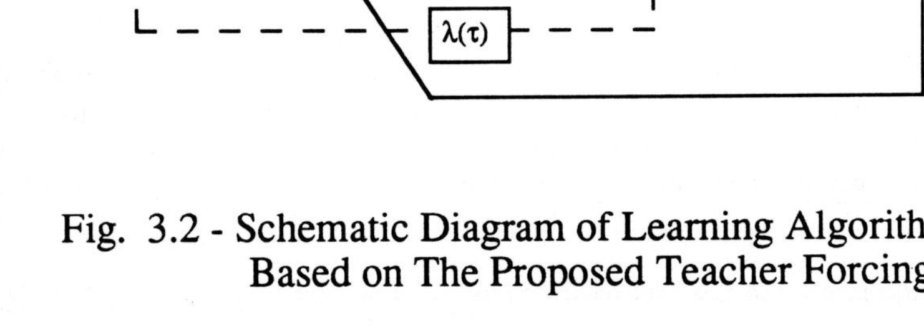

Fig. 3.2 - Schematic Diagram of Learning Algorithm Based on The Proposed Teacher Forcing

given by Eqs. (2.1), (2.2) and (2.3). However, there might be instances where the error can not be reduced to zero, implying that the teacher forcing term will not vanish as learning proceeds. Thus, a discrepancy in results between the learning and recall mode of the network should be expected. In an attempt to overcome this problem, we propose that the teacher's intervention in the learning process be decreased as learning progresses. Specifically, we suggest that λ be modulated as function of the error, according to

$$\lambda = 1 - e^{-E} \tag{3.3}$$

Numerical simulation[44] validate this choice.

4. TEMPORAL LEARNING ALGORITHM

In the previous section the problem of temporal learning has been formulated as a minimization of an error functional. Since the gradients of the functional with respect to the various parameters of the neural architecture, e.g., synaptic weights, neural gains, etc., are essential elements of the minimization process, in the past, major efforts have been devoted to the efficacy of their computation [37,40,43,47]. Learning algorithms are typically constructed by invoking Lyapunov stability arguments, i.e., by requiring that the error functional to be monotonically decreasing during learning time, τ. This translates into

$$\frac{dE}{d\tau} = \sum_{\mu=1}^{M} \frac{dE}{dp_\mu} \cdot \frac{dp_\mu}{d\tau} < 0 \tag{4.1}$$

One can always choose, with $\eta > 0$

$$\frac{dp_\mu}{d\tau} = -\eta \frac{dE}{dp_\mu} \tag{4.2}$$

which implements learning in terms of an inherently local minimization procedure. Attention should be paid to the fact that Eqs. (2.1) and (4.2) may operate on different time scales, with parameter adaptation occurring at a slower pace. Equation (4.2) implies that, in order to update a system parameter p_μ, one must evaluate the *"sensitivity" (i.e., the gradient)* of E, Eq. (2.3), with respect to p_μ. Furthermore, using Eq. (2.3) and observing that the time integral and derivative with respect to p_μ commute, one can write

$$\frac{dE}{dp_\mu} = \int_{t_o}^{t_f} \frac{dF}{dp_\mu}\, dt = \int_{t_o}^{t_f} \frac{\partial F}{\partial p_\mu}\, dt + \int_{t_o}^{t_f} \frac{\partial F}{\partial \bar{u}} \cdot \frac{\partial \bar{u}}{\partial p_\mu}\, dt \tag{4.3}$$

This sensitivity expression has two parts. The first term in the Right Hand Side (RHS) of Eq.(4.3) is called the "direct effect", and corresponds to the explicit dependence of the error functional on the system parameters. The second term in the RHS of Eq. (4.3) is referred to as the "indirect effect", and corresponds to the implicit relationship between the error functional and the system parameters via $\bar{u}$. In our learning formalism, the error functional, as defined by Eqs. (2.3) and (2.4), does not depend explicitly on the system parameters; therefore, the "direct effect" vanishes, i.e.,

$$\frac{\partial F}{\partial p_\mu} = 0 \tag{4.4}$$

Since F is known analytically (viz. Eqs. (2.3) and (2.4)), computation of $\partial F/\partial \bar{u}$ is straightforward. Indeed

$$\frac{\partial F}{\partial u_n} = -\ e_n \tag{4.5}$$

Thus, to enable evaluation of the error gradient using Eq. (4.3), the "indirect effect" matrix $\partial \bar{u}/\partial \bar{p}$ should, in principle, be computed.

Let us differentiate the activation dynamics, Eq. (2.1), including the teacher forcing, Eq. (3.1), with respect to p_μ. We observe that the time derivative and partial derivative with respect to p_μ commute. Using the shorthand notation $\partial(\cdots)/\partial p_\mu = (\cdots)_{,\mu}$ we obtain a set of equations to be referred to in the sequal as "Forward Sensitivity Equations" (FSEs):

$$\begin{cases} \dot{u}_{n,\mu} + \sum_m A_{nm}\ u_{m,\mu} = S_{n,\mu} & t > 0 \\ u_{n,\mu} = 0 & t = 0 \end{cases} \tag{4.6}$$

in which

$$A_{nm} = (\kappa_n - \gamma_n\ g'_n\ \partial I_n/\partial u_n)\delta_{nm} - \gamma_n\ g'_n\ T_{nm} \tag{4.7}$$

$$S_{n,\mu} = -u_n \delta_{p_\mu,\kappa_n} + \gamma_n g'_n \sum_m u_m\ \delta_{p_\mu,T_{nm}} + g'_n(\sum_m T_{nm} u_m + I_n)\delta_{p_\mu,\gamma_n} \tag{4.8}$$

In the above expressions, g'_n represents the derivative of g_n with respect to its arguments, δ denotes the Kronecker symbol and $S_{n,\mu}$ is defined as a nonhomogeneous "source". The source term contains all explicit derivatives of the neural activation dynamics, Eq. (2.1), with respect to the system parameters, p_μ. Hence, it is parameter dependent and its size is $(N\ *\ M)$. The initial conditions of the activation dynamics, Eq.(2.1), are excluded from the vector of system parameters $\bar{p}$. Thus, the initial conditions of the FSEs will be taken as zero. The solution of Eq. (2.1) will provide the matrix $\partial \bar{u}/\partial \bar{p}$ needed for computing the "indirect effect" contribution to the sensitivity of the error functional, as specified by Eq. (4.3). This algorithm is, essentially, similar to the scheme proposed by Williams and Zipser[47]. Computation of the gradients using the forward sensitivity formalism would require solving Eq. (4.6) M times, since the source term, $S_{n,\mu}$, explicitly depends on p_μ. This system has N equations, each of which requires multiplication and summation over N neurons. Hence, the computational complexity, measured in terms of multiply-accumulates, scales like N^2 per system parameter, per time step. Let us assume, furthermore, that the interval $[t_0, t_f]$ is discretized into L time steps. Then, the total number of multiply-accumulates scales like $N^4 L$. Clearly, such a scheme exhibits expensive scaling properties, and would not be very practical for large networks. On the other hand, since the FSEs are solved forward in time, along with the neural dynamics, the method has some inherent advantages. In particular, there is no need for a large amount of memory. Since $u_{n,\mu}$ has $N^3 + 2N^2$ components, the storage requirement scales as $O(N^3)$.

In order to reduce the computational complexity associated with the above technique for evaluating the "indirect effect" term in Eq.(4.3), an alternative approach can be considered. It is based upon the concept of adjoint operators, and eliminates the need to compute explicitly $\bar{u}_{,\mu}$. The critical feature of this approach is that a *single vector* of adjoint functions, $\bar{v}$, is obtained, by solving an N-dimensional system of equations once, not M times as previously. This vector contains all the information required for computing the sensitivities of the error functional, dE/dp_μ, with respect to *all* parameters, p_μ. The necessary and sufficient conditions for constructing adjoint equations are discussed by Cacuci[10].

It can be shown that an Adjoint System of Equations (ASEs) pertaining to the FSEs, Eq. (13), can be formally written as

$$-\dot{v}_n + \sum_m A^T_{nm}\, v_m = S^*_n \qquad t > 0 \tag{4.9}$$

where the superscript "T" denotes transposition, and the indices n and m range from 1 to N. In order to specify Eq. (4.9) in closed mathematical form , we must define the source term, S^*_n, and the initial conditions (specified at *initial* or *final time*) for the system. Both should be independent of $\bar{u}_{,\mu}$ and its derivatives.

In order to construct an expression which can be used to eliminate the dependence of the "indirect effect" term of the sensitivities on $\bar{u}_{,\mu}$, we have to form the inner product of the vectors $\bar{S}^*$ and $\bar{u}_{,\mu}$. This is done by taking the inner product of the FSEs, Eq. (4.6), by $\bar{v}^T$, and the ASEs, Eq. (4.9), by $\bar{u}^T_{,\mu}$, and by subtracting the two resulting equations. Then, integrating over the time interval $[t_o, t_f]$, yields the bilinear form:

$$[\bar{v}^T\, \bar{u}_{,\mu}]_{t_f} - [\bar{v}^T\, \bar{u}_{,\mu}]_{t_o} = \int_{t_o}^{t_f} (\bar{v}^T\, \bar{S}_{,\mu})\, dt \; - \int_{t_o}^{t_f} (\, \bar{u}^T_{,\mu}\, \bar{S}^*)\, dt \tag{4.10}$$

In the above equation $\bar{S}_{,\mu}$ denotes the inhomogeneous source term of the FSEs, Eq. (4.8), and $[\ldots]_t$ denotes the value of the expression in brackets evaluated at time t. By identifying

$$S^*_n = \partial F / \partial u_n \tag{4.11}$$

the second integral in the RHS of Eq. (4.10) is seen to become identical to the "indirect effect" term in Eq. (4.3). Selecting

$$\bar{v}(t=0) = 0. \tag{4.12}$$

as initial conditions, will allow us to integrate the resulting ASEs, i.e., Eqs.(4.9) and (4.11), forward in time. The first term in the RHS of Eq. (4.10) can be computed by using the values of $\bar{v}$ resulting from the solution of Eqs. (4.9) and (4.11) forward in time. The main difficulty resides with the evaluation of the first term in the LHS of Eq. (4.10), i.e., $[\bar{v}^T\, \bar{u}_{,\mu}]_{t_f}$. To compute it, we now introduce an auxiliary adjoint system, formally similar to Eq. (4.9):

$$-\dot{z}_n + \sum_m A^T_{nm}\, z_m = \hat{S}_n \qquad t > 0 \tag{4.13}$$

Once again, the inner product of $\bar{u}_{,\mu}$ and $\hat{\bar{S}}$ is required. The bilinear form associated with $\bar{u}_{,\mu}$ and $\bar{z}$ is obtained in a similar fashion to the derivation of Eq. (4.10). Its expression is:

$$[\bar{z}^T\, \bar{u}_{,\mu}]_{t_f} - [\bar{z}^T\, \bar{u}_{,\mu}]_{t_o} = \int_{t_o}^{t_f} (\bar{z}^T\, \bar{S}_{,\mu})\, dt - \int_{t_o}^{t_f} (\bar{u}^T_{,\mu}\, \hat{\bar{S}})\, dt \tag{4.14}$$

By choosing

$$\hat{\bar{S}} = \bar{v}(t)\delta(t - t_f) \tag{4.15}$$

the last term of Eq. (4.14) is reduced to $[\bar{v}^T\, \bar{u}_{,\mu}]_{t_f}$, which is the quantity of interest. If we furthermore select

$$\bar{z}(t_f) = 0. \tag{4.16}$$

and take into account the initial value of $\bar{u}_{,\mu}$, Eq. (4.14) yields

$$\int_{t_o}^{t_f} (\bar{u}^T_{,\mu}\, \hat{\bar{S}})dt = [\bar{v}^T\, \bar{u}_{,\mu}]_{t_f} = \int_{t_o}^{t_f} (\bar{z}^T\, \bar{S}_{,\mu})dt \tag{4.17}$$

Note that, even though we selected $\bar{z}(t_f) = 0$, we are still interested in solving the auxiliary adjoint system, Eqs. (4.13) and (4.15), forward in time. Thus, the critical issue is how to select initial condition (i.e., $\bar{z}(t_o)$), that would result in $\bar{z}(t_f) = 0$.

In the problem under consideration, the matrix A in Eq. (4.13) is time dependent, (viz Eq. (4.7)). Thus, it is practical to integrate numerically the auxiliary adjoint equations. Usually, the same numerical scheme used for Eqs. (2.1) and (4.9) should be adopted.

For illustrative purposes only, we limit the discussion in the sequel to a first order finite differences approximation, i.e., we rewrite Eqs. (4.13) and (4.15) as

$$-\frac{(\bar{z}^{l+1} - \bar{z}^{l})}{\Delta t} + [A^T]^l \bar{z}^l = 0 \qquad 0 \leq l < L \tag{4.18}$$

Where, superscript l implies that the numerical values of the quantities of interest at time step l are used, and Δt denotes the size of the integration time step. From the above equation one can easily show that

$$\bar{z}^{l+1} = B^l \cdot B^{l-1} \cdots B^1 \cdot B^0 \bar{z}(t_o) \equiv B^{l!} \bar{z}(t_o) \tag{4.19}$$

where

$$B^l = I + \Delta t \, [A^T]^l \qquad (l = 0, 1, \cdots, L-1) \tag{4.20}$$

and I denotes the identity matrix. Using a numerical approximation for the integral, the RHS of Eq. (4.17) can be recast as

$$[\bar{v}^T \, \bar{u}_{,\mu}]_{t_f} = \bar{z}^T(t_o) \, [\sum_l [B^T]^{(l-1)!} \bar{S}^l_{,\mu}] \, \Delta t \tag{4.21}$$

Algorithmically, the computation of $[\bar{v}^T \, \bar{u}_{,\mu}]_{t_f}$ can be described as follows. At each time step l, the values of the matrices $B^{(l-1)}$, and $\bar{S}^l_{,\mu}$, are calculated using Eqs. (4.7, 4.8 and 4.20). The needed value of $B^{(l-1)!}$ is computed by multiplying the stored value of $B^{(l-2)!}$ by $B^{(l-1)}$. The result is not only stored for the next iteration, but also used to calculate the product of $B^{(l-1)!}$ by $\bar{S}^l_{,\mu}$ which, in turn, is added up. The initial conditions $\bar{z}(t_o)$ can easily be found at time t_f, i.e., at iteration step L, by solving the system of linear algebraic equations:

$$B^{(L-1)!} \bar{z}(t_o) = \bar{v}(t_f) \tag{4.22}$$

To summarize, in this new method the computation of the gradients of the error functional (i.e., Eq. (4.3)), involves two stages, corresponding to the two unknown terms in Eq. (4.10). The RHS term is calculated from the adjoint functions, $\bar{v}$, obtained by integrating Eqs. (4.9) and (4.11) forward in time along with the neural activation dynamics, Eq. (2.1). The corresponding computational complexity is $O(N^2L)$. The LHS term is calculated via Eqs. (4.13) and (4.15), and involves two steps: a) kernel propagation, Eq. (4.21), which requires multiplication of two matrices B^l and $B^{(l-1)!}$ at each time step; hence, its computational complexity scales as N^3L; b) solution of the linear algebraic system (4.22) with computational complexity of $O(N^2)$. Thus, the overall computational complexity of this approach is $O(N^3L)$. Notice, however, that here the storage needed is minimal and equal to $O(N^2)$.

5. STATIC LEARNING ALGORITHM

By Analogy to the preceding section, static learning is formulated as a minimization of an error function, Eq. (2.6). Calculation of the gradients of this function with respect to the network parameters is at the core this section. By definition

$$\frac{dE}{dp_\mu} = \frac{\partial E}{\partial p_\mu} + \frac{\partial E}{\partial \bar{u}} \cdot \frac{\partial \bar{u}}{\partial p_\mu} \tag{5.1}$$

Since the error function E is known analytically, Eq. (2.6), the computation of $\partial E/\partial p_\mu$ and $\partial E/\partial \bar{u}$ is straightforward. The quantity that needs to be determined is the vector $\partial \bar{u}/\partial p_\mu$. Following the same procedure as outlined in the previous section, we differentiate the neural activation dynamics (2.1), arriving at the "Forward Sensitivity Equations" (FSEs):

$$\begin{cases} \dot{u}_{n,\mu} + \sum_m A_{nm}\, u_{m,\mu} = S_{n,\mu} & t > 0 \\ u_{n,\mu} = 0 & t = 0 \end{cases} \tag{5.2}$$

where A_{nm} and $S_{m,\mu}$ are the same as in Eqs. (4.7) and (4.8) respectively.
However, since we are here interested only in the steady state solution, Eq. (5.2) reduces to an algebraic equation. i.e.,

$$\sum_m A_{nm}\, \tilde{u}_{m,\mu} = S_{n,\mu} \tag{5.3}$$

where $\tilde{u}_{n,\mu}$ can be viewed as the steady state solution of the FSE, Eq. (5.2), and denotes the sensitivity of $\tilde{u}_n$ with respect to p_μ.

Since the source term in Eq. (5.3) explicitly depends on p_μ, computing dE/dp_μ, requires solving the above system of N algebraic equations for each parameter p_μ. Therfore, the total number of multiply-accumulate instructions scales as $O(N^4)$.The high cost of such a scheme is circumvented by resorting, once again, to adjoint operators, introduced in the previous section. The Adjoint System of Equations (ASEs) corresponding to the FSEs, Eq. (5.3), is given by

$$\sum_m A^T_{nm}\, \tilde{v}_m = S^*_n \tag{5.4}$$

By definition, for algebraic operators

$$\tilde{v}^T\, \bar{S}_{,\mu} = \tilde{u}^T_{,\mu}\, \bar{S}^* \tag{5.5}$$

If we identify

$${}^k S^*_n \equiv \frac{\partial E}{\partial\, {}^k\tilde{u}_n} = \begin{cases} {}^k\tilde{u}_n - {}^k a_n & \textbf{if } n \in S_O \\ 0 & \textbf{if } n \in S_H \cup S_I \end{cases} \tag{5.6}$$

we observe that the source term for the adjoint equations is independent of the specific parameter p_μ. Hence, the *solution of a single set of adjoint equations will provide all the information required to compute the gradient of E with respect to all parameters.* To underscore that fact we shall rewrite Eq. (5.1) as

$$\frac{dE}{dp_\mu} = \frac{\partial E}{\partial p_\mu} - \tilde{v}^T \cdot \bar{S}_{,\mu} \tag{5.7}$$

Notice that the above ASE, Eq. (5.4), is linear in ${}^k\tilde{v}$. Furthermore, it has the same mathematical characteristics as the neural activation dynamics (2.1). Its components can be obtained as the equilibrium points, (i.e., $\dot{v}_n \rightarrow 0$) of the adjoint neural dynamics

$$\dot{v}_n + \sum_m A^T_{nm}\, v_m = S^*_n \qquad t > 0 \tag{5.8}$$

As a specific example, let us conclude by deriving the learning equations for the synaptic weights, T_{nm}. Recall that

$$\frac{dE}{dp_\mu} = \frac{\partial E}{\partial p_\mu} + \sum_k {}^k\tilde{v} \cdot {}^k \bar{S}_{,\mu} \tag{5.9}$$

where ${}^k\bar{S}_{,\mu}$ is given by Eq. (4.8). Since by definition, $\partial E/\partial p_\mu = 0$, the explicit evolution of the synapse connecting the $m-$th to the $n-$th neuron is given by

$$\dot{T}_{nm} = -\eta_T \left[\sum_k \gamma^k_n \tilde{v}_n \, {}^k g'_n \, {}^k\tilde{u}_m \right] \tag{5.10}$$

It is straightforward to obtain learning equations for γ_n and κ_n in a similar fashion.

So far the adaptive learning rates, i.e., η in Eq. (4.2), have not been specified. Now we will show that, by an appropriate selection of these parameters, the convergence of the corresponding dynamical systems can be considerably improved. Without loss of generality, we shall assume $\eta_T = \eta_\kappa = \eta_\gamma = \eta$, and we shall seek η in the form[7].

$$\eta \propto |\nabla E|^{-\beta} \tag{5.11}$$

where ∇E denotes the vector with components $\nabla_T E$, $\nabla_\gamma E$ and $\nabla_\kappa E$. It is straightforward to show that

$$\frac{d}{dt} |\nabla E| = -\chi\, |\nabla E|^{1-\beta} \tag{5.12}$$

as ∇E tends to zero, where χ is an arbitrary positive constant. If we evaluate the relaxation time of the energy gradient, we find that

$$t_E = \int_{|\nabla E|_0}^{|\nabla E| \rightarrow 0} \frac{d\,|\nabla E|}{|\nabla E|^{1-\beta}} = \begin{cases} \infty & \textbf{if } \beta \leq 0 \\ \frac{1}{\beta}\, |\nabla E|_0^\beta < \infty & \textbf{if } \beta > 0 \end{cases} \tag{5.13}$$

Thus, for $\beta \leq 0$ the relaxation time is infinite, while for $\beta > 0$ it is finite. The dynamical system (5.11) suffers a qualitative change for $\beta > 0$: it loses uniqueness of solution. The equilibrium point $| \nabla E | = 0$ becomes a singular solution being intersected by all the transients, and the Lipschitz condition is violated, as one can see from

$$\frac{d}{d \mid \nabla E \mid} \left(\frac{d \mid \nabla E \mid}{dt} \right) = -\chi \mid \nabla E \mid^{-\beta} \longrightarrow -\infty \tag{5.14}$$

where $| \nabla E |$ tends to zero, while β is strictly positive. Such infinitely stable points are "terminal attractors". By analogy with our previous results we choose $\beta = 2/3$, which yields

$$\eta = \left(\sum_n \sum_m [\nabla_T E]^2_{nm} + \sum_n [\nabla_\gamma E]^2_n + \sum_n [\nabla_\kappa E]^2_n \right)^{-1/3} \tag{5.15}$$

The introduction of these adaptive time-scales dramatically improves the convergence of the corresponding learning dynamical systems.

6. IMPLEMENTATION ANALYSIS

The neural learning framework developed in the preceding sections was applied to manipulator inverse kinematics problem. But before presenting our simulation we motivate the relevance of neural networks to solving robotics problems. Space telerobots envisaged for exacting applications in unstructured and hazardous space environments, e.g., satellite servicing, space system construction and maintenance, planetary missions etc., must be able to dexterously and adaptively manipulate objects in a nonstationary task workspace. Redundancy in the design of robot manipulators has been suggested as one means to enhance their dexterity and adaptability. In contradistinction to other engineering contexts where redundancy implies fault-tolerance or superfluity, redundancy in robotics is determined relative to the task [9]. It refers to a manipulator with more than the minimum number of degrees of freedom necessary to accomplish general tasks. The major objective motivating introduction of redundancy in robot design and control is to use the additional degrees of freedom to improve performance in complex and unstructured environments. It helps overcome kinematic, mechanical and other design limitations of non-redundant manipulators. Also, the extra degrees of freedom can be used during real-time manipulator operation to simultaneously achieve end-effector trajectory control while satisfying additional constraints.

There are two primary problems in developing control strategies which take advantage of redundancy. First, given the initial and final end-effector task coordinates, simultaneously generate, in real time,y a Cartesian-space trajectory that can achieve a goal (*the path planning problem*) and a set of joint space trajectories which cause the end-effector to follow the desired trajectory (*inverse kinematics problem*) while satisfying additional constraints, such as obstacle avoidance, servo-motor torque minimization, singularity avoidance, and joint limit avoidance. Developing algorithms to use the additional degrees of freedom to satisfy constraints is known as the *redundancy resolution* problem [6,9]. Secondly, provide adaptive mechanisms for responding to any unforeseen changes in the workspace or the manipulator geometry. Despite a tremendous growth in research activity on "model-based" adaptive control algorithms, the above problems entail a level of computational and paradigmatic complexity far exceeding that what can be provided by the existing strategies. Artificial neural networks, on the other hand, as discussed previously, due to their ability to perform functional synthesis in real-time, could provide an attractive alternate basis towards designing real-time manipulator control architectures.

Manipulator kinematics addresses the problem of computing temporal evolution of joint coordinates from the motion of robot end-effector. A forward kinematic function, Φ, is a nonlinear differentiable function which uniquely relates a set of N_Q joint variables, $\bar{q}$, to a set of N_X task-space coordinates, $\bar{x}$: $\bar{x} = \bar{\Phi}(\bar{q})$. For serial chain robot manipulators the forward kinematic function is easily derived. The more difficult problem which is of primary practical interest in manipulators kinematics is the inverse problem:

$$\bar{q} \quad = \quad \Phi^{-1}(\bar{x}) \tag{6.1}$$

i.e., determine one or more sets of joint configurations which take the end-effector into a desired task position and orientation in the workspace. While the inverse kinematic function is highly nonlinear, closed form analytical solutions can be found for a number of non-redundant manipulators with special architecture. Complete positioning capability in Cartesian space can be nominally achieved using only six degrees of freedom. However, most manipulators have degenerate configurations, or kinematic singularities, near which small displacements of the end-effector requires physically unrealizable joint speeds. These singularities effectively lead to a loss of usable workspace and capability, and there is a strong incentive to design robots with additional degrees of freedom. Thus, a robot manipulator is kinematically redundant if the number of its degrees of freedom is greater than the dimension of the end-effector task space. In contradistinction to other engineering contexts, where redundancy per se, implies *fault-tolerance*, i.e., component duplication allowing for continued system functionality in the event of an element failure or *superfluity* i.e, an unneeded excess capacity, redundancy in robotics is determined relative to the task [9]. For example, a 6-DOF manipulator could be redundant with respect to tasks with symmetry about one axis, while an arm with 3 or more joints is redundant for achieving any end-effector position in a two-dimensional space. The major objective motivating introduction of redundancy in robot design and control is to use the additional degrees of freedom to improve performance in complex and unstructured environments. It helps overcome kinematic, mechanical and other design limitations of non-redundant manipulators, and simultaneously satisfy additional constraints, such as obstacle avoidance, minimization of actuator torques, singularity avoidance, providing greater dexterity, minimization of kinetic energy, improvement of some measure of manipulability, etc.

However, incorporation of redundancy injects additional complexity into the problem. For redundant manipulators, the kinematic equations relating the specified end-effector task coordinates to the unknown joint angles may not have a unique solution, and in general the problem is both ill-posed and ill-conditioned [27], i.e., solutions may not necessarily exist, and if they do exist at all, they are likely to be nonunique. Such solutions usually depend discontinuously on the input, and hence are likely to be unstable to small errors in this input. Consider Fig. 6.1., which represent the transformation, $\bar{\Phi} : \bar{q} \longrightarrow \bar{x}$. Inevitable measurement errors or signal noise can lead to data which do not lie in the range of Φ and hence not in the domain of the inverse operator Φ^{-1}; thus no solution will exist. Similarly, the transformation may not be one-to-one, so that inverse transformation does not yield a unique result. Jeffry and Rosner [27] have analyzed the unstable behavior of such systems, by examing the matrix equation $\bar{x} \;=\; \Phi\bar{q}$. If $\bar{x}$ contains errors (say observational) Δx, then a direct inverse in the least-squares sense, i.e., generalized pseudoinverse, would yield the elementary solution

$$\bar{q} \;=\; (\Phi^T\Phi)^{-1}\;\Phi^T\;[\bar{x} + \delta\bar{x}].$$

If the row vectors of $\Phi^T\Phi$ are not fully linearly independent, then $\Phi^T\Phi$ will be nearly singular, and the above inversion will greatly magnify the error, possibly driving it to dominate the solution. Therein lies the basis for the ill-conditioned nature of the problem.

Often an infinite number of joint-configurations can be obtained to satisfy a given end-effector configuration. However, it can be shown [9] that the infinity of solutions can be mapped into a finite set of manifolds. Because of this infinity of solutions, many redundant manipulator investigators have chosen to focus on the instantaneous or differential kinematics, which uses a jacobian-matrix to relate end-effector velocities to the joint velocities. The jacobian is defined as

$$\dot{\bar{x}} \;=\; \mathbf{J}(\bar{q})\;\dot{\bar{q}} \tag{6.2}$$

For redundant robots the manipulator Jacobian is not uniquely invertible, and pseudo-inverse techniques can be used to select a solution from the infinity of possible solutions in the null space of $\mathrm{J}(\bar{q})$. Eq. (6.2) is often referred to as the inverse kinematics solution, although (6.1) is the true inverse kinematics problem.

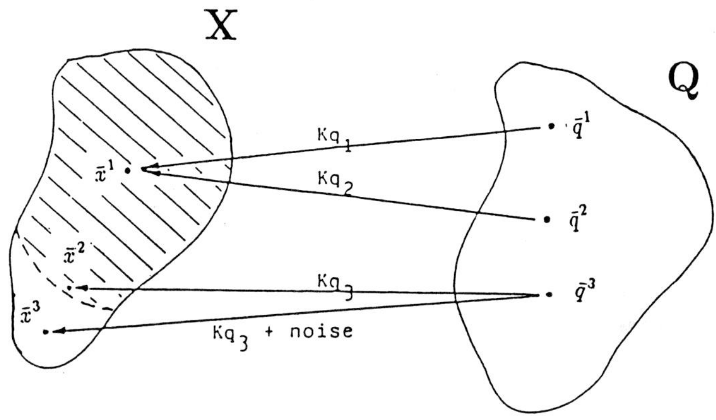

Figure 6.1 Schematic representation of inverse problems and the origin of their Ill-posed and ill-conditioned nature.

Given a desired end-effector velocity, $\dot{\bar{x}}$, the joint velocities, $\dot{\bar{q}}$ can simply be determined by:

$$\dot{\bar{q}} \;=\; \mathbf{J}^{\dagger}(\bar{q})\dot{\bar{x}} \tag{6.3}$$

where $\mathbf{J}^{\dagger}(q)$ is the pseudo-inverse, or a weighted pseudo-inverse, of the manipulator Jacobian matrix. This redundancy resolution solution minimizes a weighted quadratic norm of instantaneous joint velocities. The end-effector velocities are typically generated by a path planning algorithm, and the joint velocities computed by Eq. (6.3) are used as the reference input to a joint-space control system.

This solution can be modified by adding a null space component to the joint velocities [13], i.e.,

$$\dot{\bar{q}} \;=\; \mathbf{J}^{\dagger}(\bar{q})\dot{\bar{x}} \;+\; (I \;-\; \mathbf{J}^{\dagger}(\bar{q})\mathbf{J}(\bar{q}))\bar{z} \tag{6.4}$$

where $\bar{z}$ is an arbitrary vector. The term $(I - \mathbf{J}^{\dagger}(\bar{q})\mathbf{J}(\bar{q}))$ projects this arbitrary vector into the null space of the manipulator. Physically, any motion in the null space is an instantaneous internal motion of the manipulator which causes no motion of the end-effector. Many redundancy resolution criteria can be developed as potential functions, and $\bar{z}$ might be the gradient of the resolution potential function, i.e., , $\bar{z} = \alpha \nabla \Psi(\bar{q})$, where α is a weighting factor. Then for a given end-effector configuration, the gradient of this function is used to control joint velocity in the redundant directions, in a manner that forces the manipulator to seek an optimal configuration. However, the pseudo-inverse resolution techniques are generally not cyclic [13], i.e. these techniques do not generate closed joint-space trajectories corresponding to closed end-effector trajectories, thereby posing a serious limitation for practical implementations. Other researchers have used the null space of the jacobian, which corresponds to the self-motion of the robot, to optimize various performance criteria. For example, Liegeois [33] has developed a gradient projection scheme that utilizes the null space of the Jacobian to optimize a joint-position dependent, scalar performance criterion. However, the existing algebraic methods are, in general, very expensive computationally, and are unable to find global redundancy resolution optima with respect to multiple criteria in real-time. Also the manipulators can have more than one distinct internal motion for a given end-effector location but the instantaneous methods only optimize over one internal motion, and therefore can miss the true optimum which lies on another internal motion [9].

So in the absence of closed form solutions, off-line iterative approximation techniques based on "local-methods" have been used to solve the inverse transformation problem. In this context, Goldenberg et al. [15] have proposed an "augmented task method" that uses a modified Newton-Raphson method to simultaneously obtain all the joint variables. They partition the augmented Jacobian matrix into an invertible non-redundant component and a redundant component to obtain approximate bounds on the magnitude of the joint angles. A nonlinear constrained optimization is then performed to determine the angular displacements for the redundant joints by satisfying some auxiliary criteria. The resulting values are used to compute the Newton-Raphson correction that minimizes an error-residual between desired and current end-effector coordinates. Despite its versatility, this techniques suffers from algorithmic singularities, since it fails to ensure the non-singularity of the Jacobian-matrix partition prior to start of each iteration. Also, for a large number of degrees of freedom, the nonlinear optimization algorithm during each iteration induces a significant computational complexity.

In a significantly different approach Burdick [9] conducts a topological and geometrical analysis of the kinematics of redundant manipulators. Formulating inverse kinematics as a global manifold mapping problem, he uses the singularities of the forward kinematics to partition the configuration space manifold into disjoint regions. The topological characteristics of these regions and their forward mapping are then used to rigorously analyze kinematic properties, such as bounds nature and number of singularities that must be encountered along an arbitrary cyclic path and bounds on the number of inverse kinematic solutions. Currently formal procedures are being developed for translating this qualitative insight to quantitative algorithms that could aid the design and control of redundant manipulators.

In contrast to the algebraic and iterative strategies mentioned above, neuromorphic approaches to the inverse kinematics problem entail systems composed of many simple processors ("neurons"), fully or sparsely interconnected, whose function is determined by the topology and strength of the interconnections. The synaptic elements of such neural systems must capture the transcendental kinematic transformations using *a priori* generated examples enabling subsequent generalization to other points in the workspace. Thus, the inverse transformation equations do not need to be explicitly programmed or derived. Once they have been learned, the network's inherent self-organizing abilities enable it to adapt to changes in the environment, e.g. planning joint trajectories in the presence of obstacles, or to any unforseen changes in the mechanical structure of the manipulator, with little effort [35]. Within a neuromorphic framework, a solution of the inverse kinematic involves two phases, a training phase and a recall phase. The training phase involves encoding the inverse mapping in the network's synaptic weight space, through repeated presentations of a finite set of *a priori* generated examples, linking cartesian space end-effector coordinates to the corresponding joint angles. Once the network has acquired the nonlinear mapping imbedded within the training set, it can be used to rapidly recall, or generalize the joint configuration corresponding to any given cartesian-space orientation within it's workspace of training, thereby eliminating the computational overheads associated with the existing iterative techniques. Also, once the training cycle is completed, the time required to obtain a solution depends in a weak fashion on the number of degrees of freedom.

In the past, Josin [29], Guez [19] and Tawel et al. [42] have applied this generic neuromorphic paradigm to the inverse kinematics problem for a 3-DOF redundant manipulator. In particular, they train a heteroassociative, multi-layered feed-forward neural network using the backpropagation algorithm. The following principle is commonly used during the training process. When the system produces a wrong output on presentation of an I/O pair, the learning update rule simply changes each weight in the direction which makes the size of the error decrease as rapidly as possible. The components of this steepest descent direction in weight space are evaluated by using the chain rule to compute the partial derivatives of an error function with respect to each weight. The implementation of this weight change requires recursively propagating an error signal backward through the network, changing weights that had a large effect on the output more than those that did not. This process is repeated until the residual error between the network and target output, over all patterns, falls below a minimum acceptable tolerance.

Despite its conceptual simplicity, there are a number of non-trivial issues, both from the kinematics perspective and from the computational cost perspective that have hitherto limited the efficacy of such neuromorphic solutions to the inverse kinematics problem for redundant motion control. The major limitations, as discerned from the existing implementations, include an unacceptably large number of training iterations ($O(10^6)$) even for generalizing over small manifolds, see Tawel et al [42]. Also the interpolated angular coordinates have relatively poor precision as compared to their algebraic or iterative counterparts. Besides, the backpropagation algorithm fails to efficiently scale-up to configurations with large number of degrees of freedom. For example manipulators with seven or more degrees of freedom could not be satisfactorily trained using the standard back-propagation algorithm even after several million iterations. Furthermore the back-propagation algorithm *per se* does not provide any intrinsic mechanism to simultaneously exploit redundancy to increase the task workspace (design constraints) and satisfy additional requirements inherent to operations in an unstructured environment such as obstacle avoidance in real-time.

6.2 Implementation Results

The neurodynamical learning framework developed in this article was applied to a planar 3-degree of freedom redundant manipulator and to a spatial 7-degree of freedom human-arm like manipulator. Though either of the manipulators encompasses configurations which exhibit sufficiently the problematic complexity presented by the inverse kinematics mapping, we experimented with both examples for a variety of reasons. The 3-DOF planar manipulator was primarily used to ascertain the algorithmic correctness of our terminal teacher forcing based neural learning algorithm. It provided benchmarks for comparison with the existing backpropagation based neural network solutions [19,42], in terms of number of training iterations and training samples needed to stabilize the network, estimates on the number of synaptic elements or neurons required to successfully capture the inverse mapping and the accuracy of recalled or "generalized" joint configurations corresponding to the input end-effector coordinates. Though backpropagation-based neural networks can be applied to planar redundant manipulators with 3-DOF, they failed to scaleup to cases involving seven or more joints. Their practical applicability would thus appear to be severely limited for kinematic control of most industrial robot manipulators. We also empirically analyzed the robustness and computational speed of our artificial neural system on a 7-DOF redundant manipulator to illustrate its efficacy as a viable, real-time alternative to the existing quantitative techniques. To provide a context for the subsequent analysis we precede our discussion on the simulation results with a brief description of each candidate manipulator.

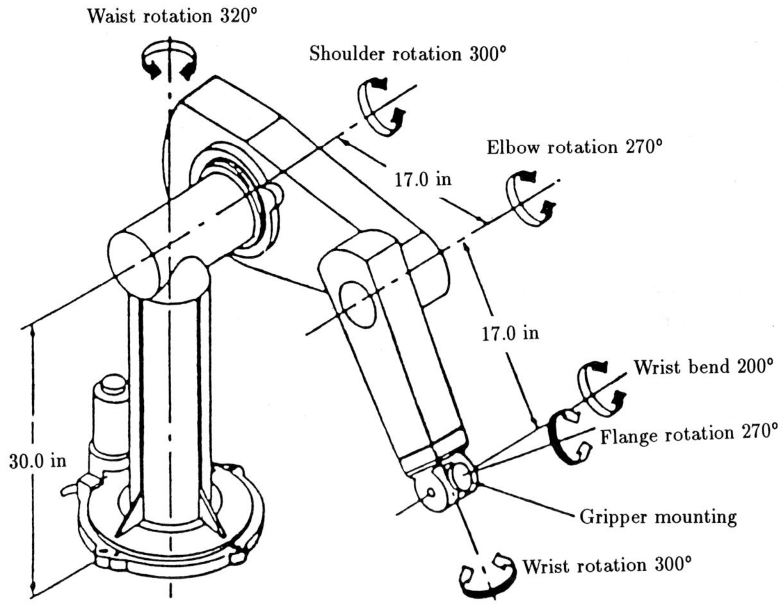

Figure 6.2 (a) Arm configuration for a PUMA 560 Robot.

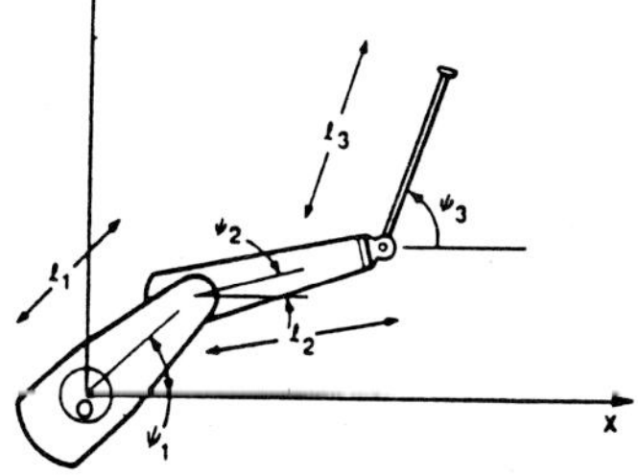

Figure 6.2 (b) PUMA 560 Robot Arm constrained to coplanar configurations.

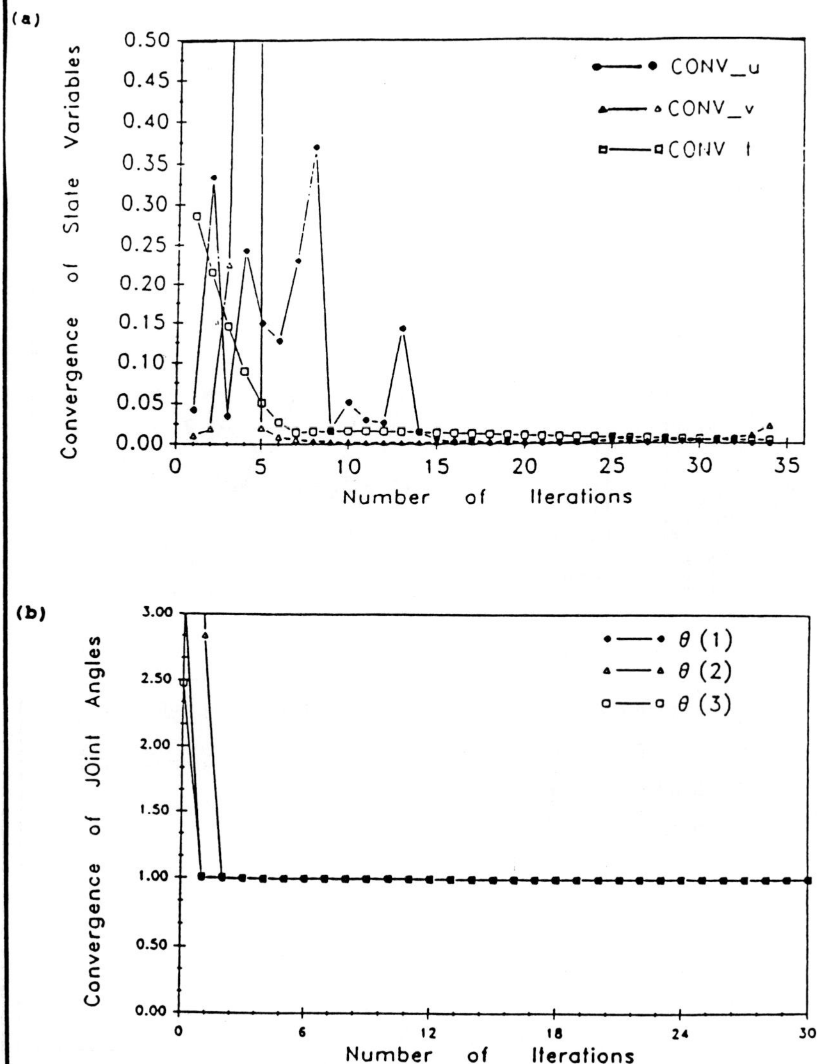

Figure 6.3 (a) Convergence of state variables; (b) Convergence of joint angles during training of 3-DOF constrained PUMA 560 Robot Arm.

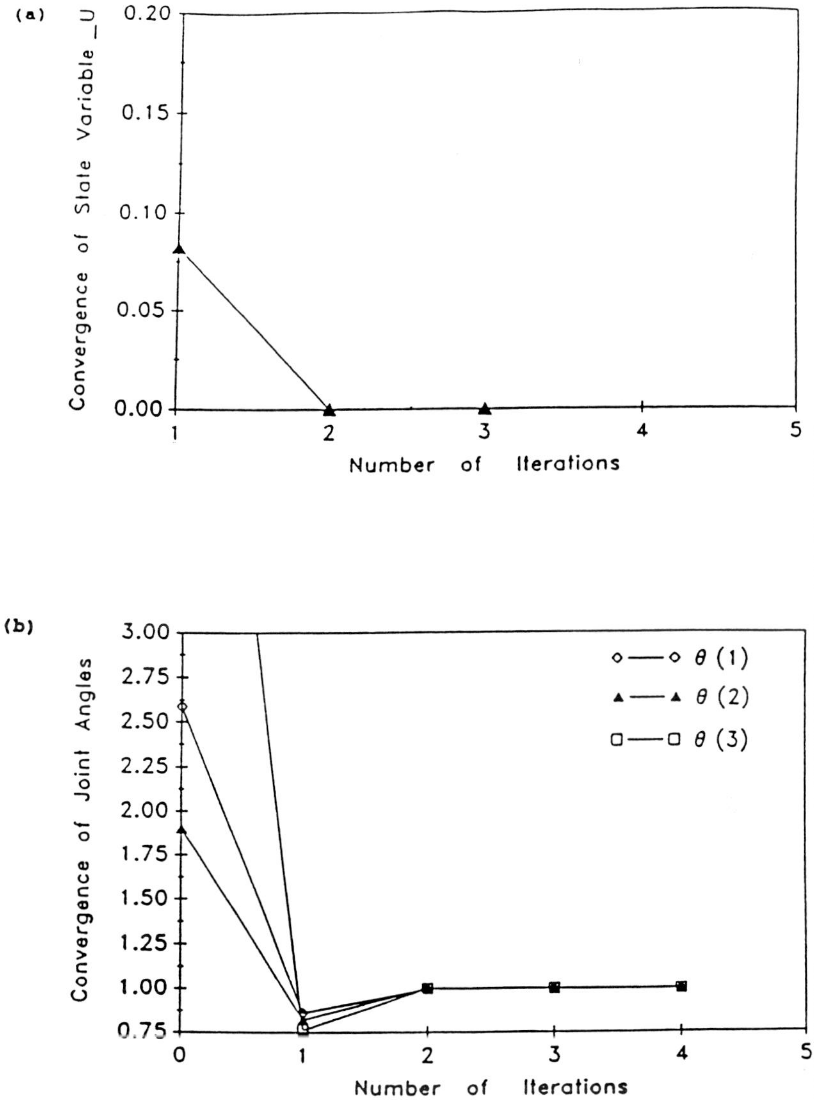

Figure 6.4 (a) Network activity; (b) Normalized behavior of output neurons during recall, in the case of 3-DOF constrained PUMA 560 Robot Arm.

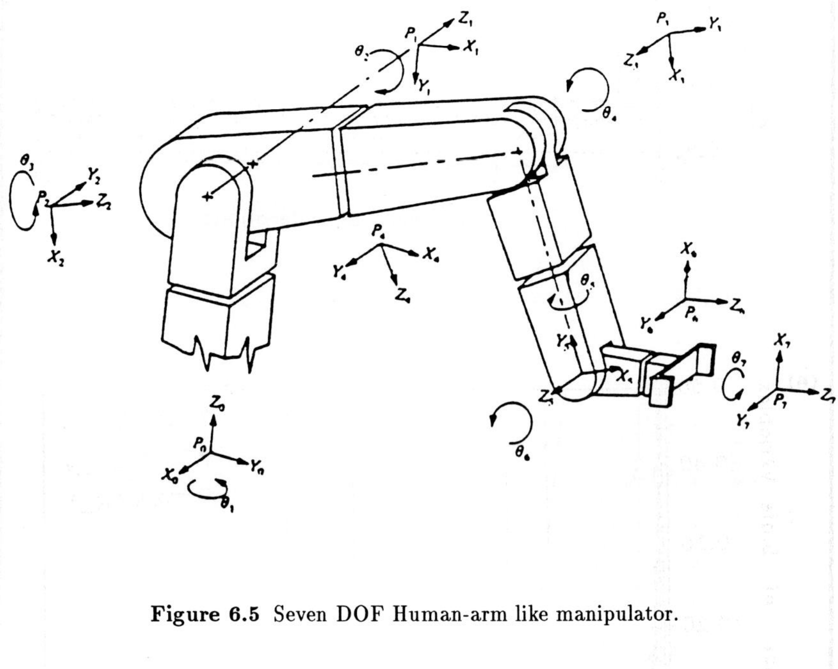

Figure 6.5 Seven DOF Human-arm like manipulator.

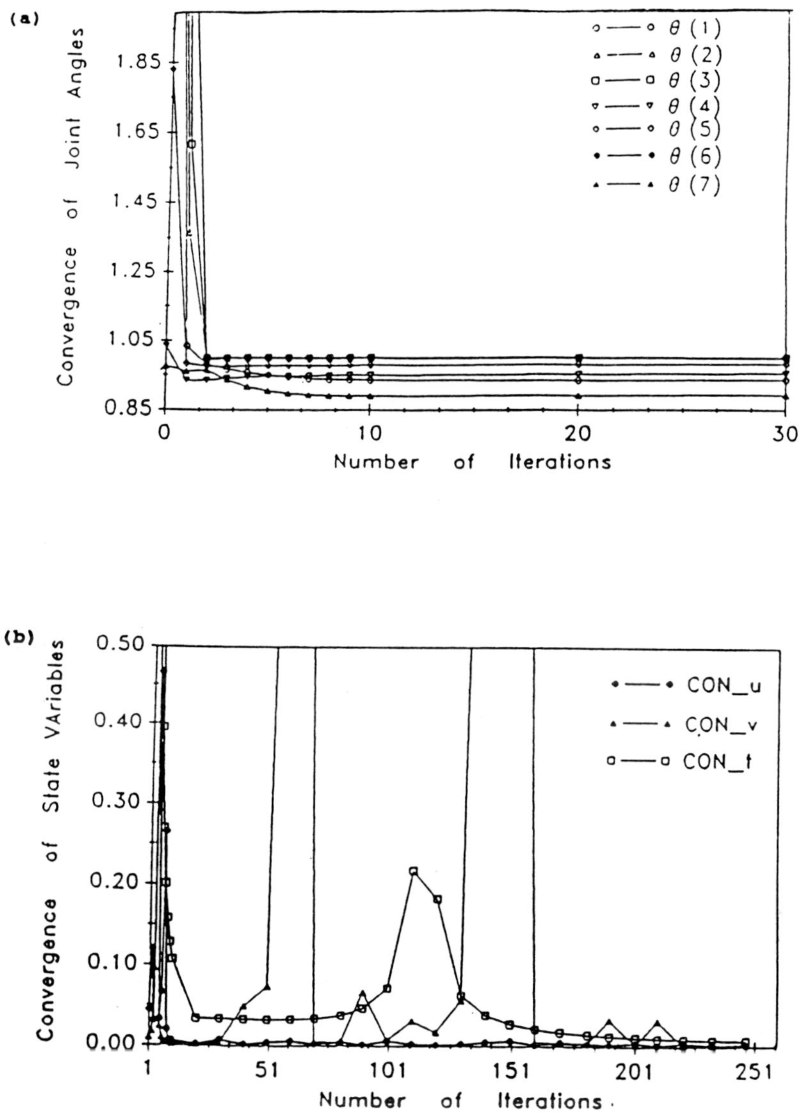

Figure 6.6 (a) Normalized behavior of state variables; (b) Convergence of joint angles, during training phase for a 7-DOF redundant manipulator.

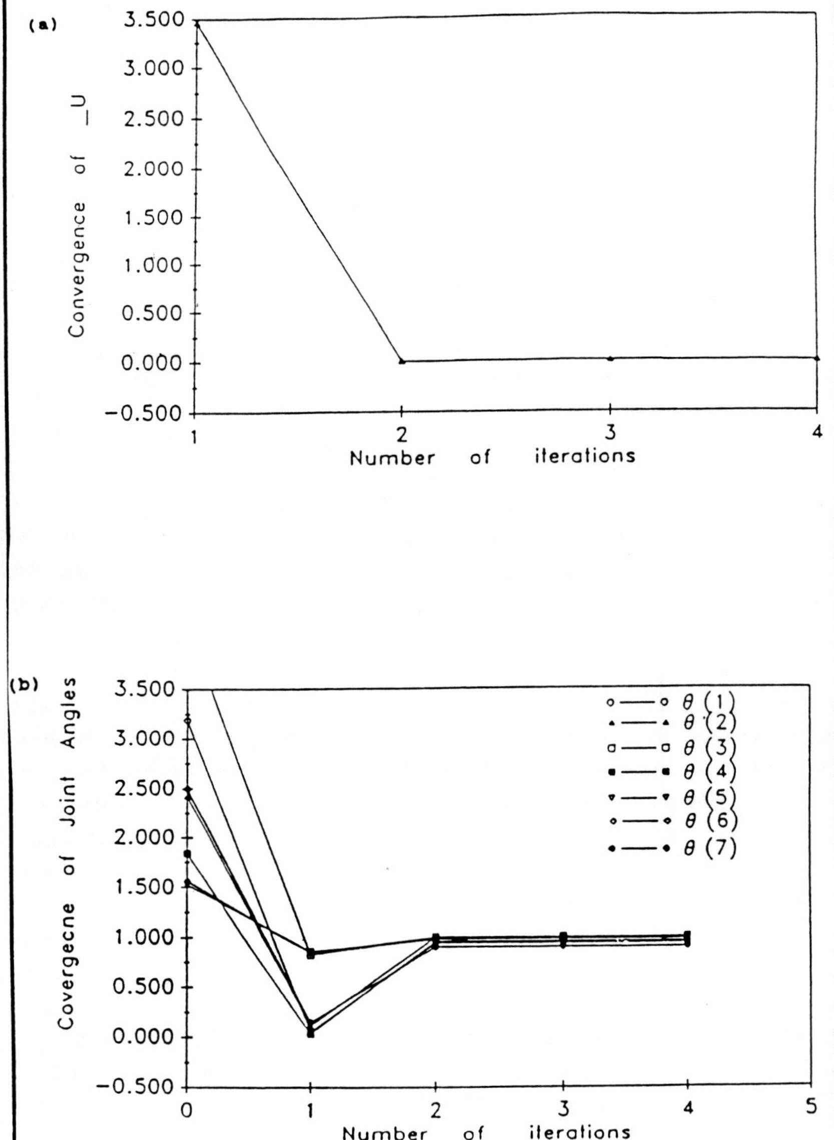

Figure 6.7 (a) Convergence of network activity; (b) Convergence of joint angles, during operational phase for a 7-DOF redundant manipulator.

The continuous-time, dynamical training procedure was simulated using parameters corresponding to a constrained configuration of the six-jointed PUMA 560 industrial robot [41]. By suppressing the motion of the shoulder, elbow and the wrist joints, the PUMA robot arm, Fig. 6.2(a), was restricted to motion in a vertical plane only as shown in Fig. 6.2(b). Fig. 6.3(a) illustrates the worst-case normalized behavior of the state variables and the synaptic elements during the learning phase, as the neural network acquires the inverse mapping. Figure 6.3(b) shows the convergence of the output neurons to the presented attractors during the learning phase for a particular sample. When learning has stabilized over all the training samples the network switches to its operational mode. Figure 6.4(a) displays the normalized convergence behavior of the neuronal activity, $\bar{u}$ as the network generalizes the joint angles in response to arbitrary cartesian inputs. Figure 6.4(b) illustrates the convergence of interpolated joint angles. Notice the rapid rate of convergence in computing the joint configuration as juxtaposed to conventional techniques.

In addition, the learning algorithm was applied to train the inverse kinematics transformations for Hemami's simplified 7-DOF manipulator formulation [21] of the human-arm. As shown in Figure 6.5, the following joint motions are available: the joint θ_1 provides back and forth motion about the shoulder, θ_2 provides effector elevation in the vertical plane, θ_3 enables rotation around the upper-arm axis, while θ_4 provides the elbow motion, θ_5 is around the forearm axis and θ_6 and θ_7 lead to the pitch and yaw motions of the wrist, respectively. Details on the geometric parameters, namely link length, twist angle, joint limits and offsets may be found in [21]. Forward kinematics equations, were used to generate training samples of end-effector and joint-space coordinates over the workspace volume of the robot. Figure 6.6(a) displays the normalized, worst-case temporal behavior of state variables $\bar{u}$, i.e., $\max_{\forall\, n} \mid (u_n^{t+1} - u_n^t)/u_n^t \mid$, adjoint variables $\bar{v}$ and the synaptic elements, $\hat{T}$, during the learning phase. Figure 6.6(b) shows the variation in activity at the output neurons during the learning cycle corresponding to one of the training pairs being presented to the network. Note that the system learns the inverse mapping in a few hundred iterations only, as compared to the several million iterations required by the gradient descent-based backpropagation algorithm. The computational efficacy of our neuromorphic learning algorithm may be estimated from the rate of variation in network activity during the operational phase of the network as shown in figures 6.7(a) and 6.7(b) As depicted in Fig. 6.7(a), once the network has acquired the inverse transformations, it may be used to recall or generate the joint angles needed to achieve any arbitrary end-effector coordinate, within the workspace of the manipulator, in very few dynamical iterations.

7. CONCLUSIONS

This paper has addressed a fundamental problem in computational AI – developing a new class of massively parallel, neural learning algorithms for robustly abstracting complex nonlinear transformations, e.g., functional, spatial, temporal and statistical invariants, from representative samples, in real-time. Provision of such a capability is at the core of many difficult problems in robotics, signal processing, remote sensing, control, etc. In contradistinction to existing dynamical neural learning formalisms, our models encode information as singular, rather than regular, solutions to neurodynamics equations using the notion of terminal attractors. The infinite local stability resulting from violation of Lipschitz conditions, enables dramatic speedups during the learning process. Our notion of real-time pertains to operational responses that can be obtained in a few time constants of the individual neurons.

We exploited a powerful tool for sensitivity analysis of nonlinear systems to put on a firm mathematical basis our results in computational learning. We provide a formal framework for global computation of sensitivities. In particular the concept of *adjoint operators* was used to enable computation of changes in the network's response due to perturbations in all system parameters, using the solution of a single set of appropriately constructed linear equations. The powerful repertoire of adaptive neural learning formalisms presented in this paper were then exploited to provide an enabling core for addressing a fundamental problem in robotics - the design of autonomous robots designed to perform tasks in unstructured and unpredictable environments.

ACKNOWLEDGMENTS

This research was carried out at the Center for Space Microelectronics Technology, Jet Propulsion Laboratory, California Institute of Technology. Support for the work came from agencies of the U.S. Department of Defense, including the Naval Weapons Center (China Lake, CA), and from the Office of Basic Energy Sciences of the U.S. Department of Energy, through an agreement with the National Aeronautics and Space Administration. The authors acknowledge helpful discussions with M. Zak.

References

[1] Amari, S.-I. (1979). "A Neural Theory of Association and Concept Formation", *Bio. Cyber.*, 26, 175-185.

[2] Amari, S.-I. (1983). "Field Theory of Self-organizing Neural Networks," *IEEE Trans. Sys., Man, Cyber.*, SMC-13(5), 741-748.

[3] Amari, S.-I. and M.A. Arbib (Eds.) (1982). *Competition and Cooperation in Neural Networks*, New York: Springer-Verlag.

[4] Anderson, J.A. (1988). *Neurocomputing*, Cambridge, MA: The MIT Press.

[5] Arbib, M.A. (1989). *The Metaphorical Brain*, New York, NY: John Wiley.

[6] Barhen, J., N. Toomarian and S. Gulati (1990). " Application of Adjoint Operators to Neural Learning", *Applied Math Letters*, 3(3), 13-18. See also J. Barhen, S. Gulati, and M.Zak, Neurodynamics of Redundant Robots: I. Manipulator Inverse Kinematics", *First Annual Int'l Neural Network Society Meeting*, Boston, MA, 1989.

[7] Barhen, J., S. Gulati and M. Zak (1989). "Neural Learning of Constrained Nonlinear Transformations", *IEEE Computer*, 22(6), 67-76.

[8] Baron, R.J. (1987). *The Cerebral Computer*, Hillsdale, NJ: Lawrence Erlbaum Associates.

[9] Burdick, J.W. (1988). " Kinematic Analysis and Design of Redundant Robot Manipulators," *Ph. D. Thesis*, Dept. of Mech. Engg., Stanford University.

[10] Cacuci, D.G. (1981). "Sensitivity theory for nonlinear systems". *Journal Math. Phys.*, **22** (12), 2794-2802.

[11] Carpentar, G.A. and S. Grossberg (1987). "A Massively Parallel Architecture for a Self-Organizing Neural Pattern Recognition Machine", *Comp. Vis., Gra., and Image Proc.g*, 37, 54-115.

[12] Carpentar, G.A. and S. Grossberg (1987). "ART 2: Self-Organization of Stable Category Recognition Codes for Analog Input Patterns", *Applied Optics*, 26(23), 4919-4930.

[13] Dubey, R.V., J.A. Euler and S.M. Babcock (1988). "An Efficient Gradient Projection Optimization Scheme for a 7-DOF Redundant Robot with a Spherical Wrist ", in *Proc. IEEE Int'l. Conf. on Rob. and Aut.*, Philadelphia, I, 28-36.

[14] Golden, R.M. (1986). "The 'Brain-state-in-a-box' Neural Model is a Gradient Descent Algorithm", *Jour. of Math. Psy.*, 30, 73-80.

[15] Goldenberg, A.A. (1985). "A Complete Generalized Solution to the Inverse Kinematics of Robots," *IEEE Jour. of Rob. and Aut.*, RA-1(1), 14-20.

[16] Grossberg, S. (1987a). *The Adaptive Brain, I: Cognition, Learning , Reinforcement and Rhythm*, Amsterdam: Elsevier/North-Holland.

[17] Grossberg, S. (1987b). *The Adaptive Brain, II: Vision, Speech, Language and Motor-Control*, Amsterdam: Elsevier/North-Holland.

[18] Grossberg, S. (1988). "Nonlinear Neural Networks: Principles, Mechanisms, and Architectures," *Neural Networks*, 1(1), 17-62.

[19] Guez A. (1988), "Solution to the Inverse Kinematics Problem in Robotics by Neural Networks," in *Proc. of 2nd Int'l Conf. on Neural Networks*, San Diego, CA, 2, 617-624.

[20] Hecht-Nielson R., "Neurocomputer Applications in Performance Limits of Optical, Electro-Optical and Electronic Artificial Neural System Processors" in Proc. SPIE,

634, 277.

[21] Hemami A. (1987). "On a Human-Arm Like Mechanical Manipulator", *Robotica*, 5, 23-28.

[22] Hinton, G.E. (1987). "Connectionist Learning Procedures", *Tech. Rep. CMU-CS-87-115*, Comp. Sci. Dept., CMU, Pittsburg, PA.

[23] Hopfield, J.J. (1982). "Neural Networks and Physical Systems with Emergent Collective Computational Abilities", *Proc. Nat'l. Acad. Sci.* 79, 2554-2558.

[24] Hopfield, J.J. and Tank, D.W. (1985). "Neural Computation and Constraint Satisfaction Problems and the traveling Salesman", *Biol. Cyber.*, 52, 141-152.

[25] Ito, M., (1984), "The Cerebellum and Neural Control", New York: Raven.

[26] Ito, M. (Ed.) (1989). *Neural Programming*, Basel: Karger.

[27] Jeffrey, W. and R. Rosner (1986). "Optimization Algorithms: Simulated Annealing and Neural Network Processing", *The Astrophysical Journal*, 310, 473-481.

[28] Jorgensen, C. and C. Matheus (1986). "Catching Knowledge in Neural Nets", *AI Expert*, 86, 31-40.

[29] Josin G. (1988), "Neural-Space Generalization of a Topological Transformation", *Biological Cybernetics*, 59, 283-290.

[30] Kohonen, T. (1988), "An Introduction to Neural Computing", *Neural Networks*, 1(1), 3-16.

[31] Lapedes, A. and Farber, R. (1987). "Nonlinear Signal Processing using Neural Networks: Predictions and Systems Modeling", *LA-UR87-2662*, Los Alamos National Laboratory.

[32] Le Cun, Y. (1985). "A Learning Scheme for Asymmetric Threshold Networks", *Proc. Congitiva 85,* 599-607.

[33] Liegeois, A. (1977). "Automatic Supervisory Control of the Configuration and Behavior of Multibody Mechanisms", *IEEE Trans. Sys., Man and Cyber.*, SMC-7Z(12), 868-871.

[34] Llinas, R.R., (1988). *The Biology of the Brain: From Neurons to Networks*, New York, NY: W.H. Freeman and Co.

[35] Maciejewski, A.A. and C.A. Klein, (1987). "Obstacle Avoidance for Kinematically Redundant Manipulators in Dynamically Varying Environments", *Int'l Jour. on Rob. Res.*, 4(3), 109-117.

[36] Mead, C. (1989). *Analog VLSI and Neural Systems*, Reading, MA: Addison-Wesley.

[37] Pearlmutter, B.A. (1990). "Dynamic recurrent neural networks". Technical Report CMU-CS-90-196, School of Computer Science, Carnegie Mellon University, Pittsburgh, PA.

[38] Psaltis, D. and N.H. Farhat (1985). "Optical Information Processing Based on an Associative-Memory Model of Neural Networks with Thresholding and Feedback,"

Optical Letters, 10, 98-100.

[39] Rumelhart, D.E., McClelland, J.L. and the PDP Research Group (1986). *Parallel Distributed Processing*, Vol. I and II, Cambridge, MA: The MIT Press.

[40] Sato, M. (1990). "A real time learning algorithm for recurrent analog neural networks". *Biological Cybernetics*, **62** (3), 237-242.

[41] Seraji H. (1987). "Direct Adaptive Control of Manipulators in Cartesian Space", *Jour. of Rob. Sys.*, 4(1), 157-178.

[42] Tawel, R., S. Everhardt and A.P. Thakoor (1988). "Neural Networks For Robotic Control," in *Proc. Conf. on Neural Networks for Computing*, Snowbird, Utah.

[43] Toomarian, N. and Barhen, J. (1991). "Adjoint operators and non-adiabatic algorithms in neural networks". *Applied Mathematical Letters*, **4** (2), 69-73.

[44] Toomarian, N. and Barhen, J. (1992). "Learning a Trajectory Using Adjoint Functions and Teacher Forcing", *Neural Networks*, **5** (3), in press

[45] Toomarian, N. (1992). "Pattern Classification Using the Phase Velocity Approach", *Pattern Recognition Letters*, in press

[46] Waltz, D.L. (1988). "Prospects for Building Truly Intelligent Machines", *Daedalus.*

[47] Williams, R.J., and Zipser, D. (1989). "A learning algorithm for continually running fully recurrent neural networks", *Neural Computation*, bf 1 (2), 270-280.

[48] Yariv A., C.F. Neugebauer and A. Agranat, US Patent Application 07-552-772 (filed 1990). See Also A. Agranat, C.F. Neugebauer, A. Yariv, "The CCD Neural Processor,", *IEEE Trans. Cir. Syst.*, 37, 1073-1075, 1990.

[49] Zak, M. (1989). "Terminal attractors in neural networks". *Neural Networks*, **2** (4), 259-274.

